Organisms and Artifa

Life and Mind: Philosophical Issues in Biology and Psychology
Kim Sterelny and Robert A. Wilson, editors

Cycles of Contigency: Developmental Systems and Evolution
Susan Oyama, Paul E. Griffiths, and Russell D. Gray, editors, 2000

Coherence in Thought and Action
Paul Thagard, 2000

Evolution and Learning: The Baldwin Effect Reconsidered
Bruce H. Weber and David J. Depew, 2003

Seeing and Visualizing: It's Not What You Think
Zenon Pylyshyn, 2003

The Mind Incarnate
Lawrence A. Shapiro, 2004

Organisms and Artifacts: Design in Nature and Elsewhere
Tim Lewens, 2004

Organisms and Artifacts

Design in Nature and Elsewhere

Tim Lewens

A Bradford Book
The MIT Press
Cambridge, Massachusetts
London, England

First MIT Press paperback edition, 2005

MIT Press books may be purchased at special quantity discounts for business or sales promotional use. For information, please email <special_sales@mitpress.mit.edu or write to Special Sales Department, The MIT Press, 55 Hayward Street, Cambridge, MA 02142.

This book was set in Sabon by Interactive Composition Corporation and was printed and bound in the United States of America.

Library of Congress Cataloging-in-Publication Data

Lewens, Tim.
Organisms and artifacts : design in nature and elsewhere / Tim Lewens.
p. cm. — (Life and mind)
"A Bradford Book."
Includes bibliographical references and index.
ISBN 0-262-12261-8 (hc: alk. paper), 0-262-62199-1 (pb)
1. Biology—Philosophy. I. Title. II. Series.

QH331.L533 2004
570′.1—dc22

2003061768

10 9 8 7 6 5 4 3 2

Time out of mind it has been by way of the "final cause," by the teleological concept of end, of purpose or of "design," in one of its many forms (for its moods are many), that men have been chiefly wont to explain the phenomena of the living world, and it will be so while men have eyes to see and ears to hear withal.

—D'Arcy Thompson, *On Growth and Form*

It [teleology] is important, but my sense ... is that there is a feeling that basically the subject is worked out. Natural selection produces design-like objects and so function talk is appropriate.... Of course, as always in philosophy there is scope for all those bizarre counter-examples to which we all seem so addicted ...; but frankly this is the stuff of PhD theses and not the real world.

—Michael Ruse, "Booknotes," *Biology and Philosophy*

Contents

Preface

An outsider who looks at evolutionary biologists' language might think they are behind the times. What is all this talk of solutions adopted by species to deal with the problems laid down by environments? Why do biologists persist in asking what the peacock's tail, or the earwig's second penis, are for? Shouldn't they have stopped talking about the purpose of the panda's thumb over a hundred years ago? All this talk smacks of intelligent design—of artifacts, not of organisms. Yet Darwin taught us (or maybe it was Hume) that organisms are not artifacts. What is more, it is often the biologists keenest to distance themselves from any nonsense about intelligent design who are nonetheless the first to trumpet the excellence of design in nature, and who look most eagerly for functional explanations for any and all organic traits or behaviors. Could it be that there are vestiges of natural theology lurking in this language of design? This book is an investigation of an analogy—the analogy between the processes of evolution and the processes by which artifacts are created. I try to show how looking at the two domains together can shed light on how we should explain the form of both organic and inorganic objects, and how our conclusions about natural design can inform various philosophical projects. It is important to understand the organism/artifact analogy for what it can tell us about biology, technology, and philosophy.

This book is addressed primarily at philosophers of biology; this said, I hope also that real biologists, students of technology, philosophers of mind—even some civilians—will find lots to interest them. Chapter 2 looks at biology alone, yet its conclusions are used as foundations for work done in all the remaining chapters, and all types of reader are encouraged to look at it. Biologists will perhaps be most interested in this second chapter, and also in chapters 3 and 4, in which I discuss how the design-based thinking of adaptationism can lead us astray, how selection explains adaptation, and how selection explanations and developmental explanations might come into conflict. Philosophers of mind will get most from chapters 5 and 6, where I approach the old-fashioned question of how to understand functions, and I put forward a nonhistorical, deflationary, analysis of biological function. Historians of technology, and perhaps other students of the made world, may be most interested in the final chapter, where I look to the prospects for an informative evolutionary model of technological change. Finally, a thread runs through the book that offers an explanation that does not itself look to intelligence as the justification for the appearance in biology of a vocabulary that is saturated with intelligence and intention. This explanation closes off one set of routes to intelligent design creationism, and readers interested in that debate should also find this work useful.

Parts of the book, or at least some of its arguments, have appeared elsewhere. Chapter 1 draws from "No End to Function Talk" (*Studies in History and Philosophy of Biological and Biomedical Sciences* 32, 2001) and "Function Talk and the Artifact Model" (*Studies in History and Philosophy of Biological and Biomedical Sciences* 31, 2000). Chapter 2 borrows some arguments from a paper I coauthored with Denis Walsh and André Ariew called "The Trials of Life" (*Philosophy of Science* 69, 2002). Chapter 3 is an augmented version of "Adaptationism and Engineering" (*Biology and Philosophy* 17, 2002), and I am grateful to Kluwer Academic Publishers for permission to reproduce material from that article here. Chapter 7 includes some arguments that were first aired in "Darwinnovation!" (*Studies in History and Philosophy of Science* 33, 2002).

There has been a good deal published in the last couple of years on the topics addressed in this book, and I have not been able to take account of all of it. One book—Peter McLaughlin's (2001) *What Functions Explain*—came to my attention only as the final draft was being prepared. I have tried to give brief indications in chapter 5 of some of the points over which we agree and differ, but I have not had time to integrate a discussion of McLaughlin's views in a more comprehensive way.

There are many people without whose help this book would have been even worse than it is. My first and greatest debt is to Nick Jardine, who was my supervisor when *Organisms and Artifacts* was a Ph.D. dissertation. Nick is an ideal supervisor: always encouraging, always willing to make time to read work, and full of the most helpful and insightful comments.

The Department of History and Philosophy of Science in Cambridge is a wonderful place to have worked over the last five years. I am grateful to Joanna Ball, Anjan Chakravartty, Jim Endersby, Marina Frasca-Spada, Anandi Hattiangadi, Tamara Hug, Martin Kusch, Peter Lipton, Helen Macdonald, Neil Manson, Hugh Mellor, Greg Radick, Matthew Ratcliffe, David Thompson, and Jill Whitelock for various combinations of advice, criticism, support, and friendship.

A large proportion of an early draft was written during the academic year 1999 to 2000, when I was a visiting student in the Centre for the Philosophy of Natural and Social Sciences at the London School of Economics. At the LSE I would like to thank Helena Cronin, Oliver Curry, Dylan Evans, Nicholas Humphrey, and Richard Webb for comments on talks I gave there.

Many people outside Cambridge and the LSE have been exceptionally generous in providing comments on my work and assorted ideas. André Ariew, David Buller, Paul Sheldon Davies, John Dupré, Peter Godfrey-Smith, Paul Griffiths, Richard Lewontin, Mohan Matthen, Joel Mokyr, Elliott Sober, and Chris Stephens all deserve thanks. Denis Walsh merits special mention for comment and correspondence beyond the call of duty. I'm grateful to Kim Sterelny and Rob Wilson for taking on the manuscript at the MIT Press, and especially to Rob Wilson and an anonymous reader from MIT for some very helpful comments on the penultimate draft.

For financial support I am grateful to the Arts and Humanities Research Board, to the Raymond and Edith Williamson Fund, to Corpus Christi College, and to Clare College. My mum and my sister kindly helped with proofreading over Christmas dinner. Christina McLeish did a great job on the index. I also owe personal debts to Emma Gilby, Cesare Hall, Annette van der Kolk, and Emily Roche.

1

Meaning and the Means to an Understanding of Ends

1.1 Design in Nature

Biology is unique among the natural sciences in its use of a family of concepts that might seem better suited to the description and explanation of artifacts than the description and explanation of organisms. Artifacts are objects made by intelligent agents; organisms—most of them, at least—owe their construction to no agent. When we think about artifacts of all kinds—shoes, ships, sealing wax—we find it natural to ask what might be their functions, and the functions of their parts, what problems they were made to solve, and so forth. Biologists, and evolutionary biologists in particular, use a similar vocabulary when they describe and approach the organic world. They ask what the function of the stiff-legged jumping behavior (called "stotting") of Thompson's gazelles might be; they conjecture that the bony plates on the back of *Stegosaurus* had the purpose of regulating heat; they suggest that the fragile second penises of male earwigs snap off inside the vagina in order to prevent fertilization from other males; they ask what evolutionary problems our hominid ancestors might have faced in the Pleistocene, and what solutions our species might have found to meet them.

The vocabulary of intelligent design—the vocabulary of problems, solutions, purpose, and function—might seem to presuppose the existence of an intelligent designer. The human sciences may speak of purposes and problems addressed by social institutions; that is no surprise, for the human sciences range over systems that contain intelligent designers. The physical sciences generally admit no intelligent designer into their worldview; correspondingly, physicists do not speak of the purposes of electrons, and chemists do not ask what benzene rings were designed for. Thus biology is in an awkward position: it makes free with a vocabulary of design, even though modern biology recognizes no intelligent designer as the artificer of the species.

In summary, many biologists adopt what I call the *artifact model* of nature: they talk of organisms as though they were designed objects. An examination of the artifact model answers questions that are of as much interest to biologists, students of technology, and philosophers of mind as they are to philosophers keen to understand biological explanation. Much of the debate over adaptationism, for example, has been framed as a question of whether it is right to assume that organisms can be divided into traits each with its own function, in the same way that we might try to draw an exploded diagram of a car that assigns discrete functions to its parts. Developmental biologists have argued that a focus on function has led us to ignore some of the most important factors affecting form. More broadly, the investigation of the analogy between evolution and the design process has been thought by some to yield important insights regarding changes in technology itself. And philosophers of mind have thought that an account of how hearts can be supposed to pump blood, even though they may fail to do so, could yield a wholly unmysterious account of how, for example, beliefs can be supposed to represent cows, even though such a belief may fail to represent its object accurately.

This book addresses what I take to be the most pressing questions raised by the phenomena of artifact talk in biology. Such questions include: what explains the ability of biologists to use such a vocabulary? Are the terms they use mere metaphors that trade on superficial similarities between the appearance of organisms and artifacts, or are there close analogies between the processes that go into the construction of each? Might we be misled by approaching organisms as though they are collections of more or less well-designed solutions to environmental problems? Can such a framework give us a strong predictive engine for the generation of hypotheses about the workings of plants and animals—even of the human mind? Might the kinds of norms that we appear committed to—in speaking of what traits are *supposed to do*—be appropriate to solve problems in the philosophy of mind? How should we explain the appearance of artifact talk in biology and its absence in chemistry and physics? Can the function and design of artifacts themselves be approached from an evolutionary perspective? Most recent work in this area has been concerned with giving an analysis of the concept of biological function as it appears in biological journals. This is certainly an important job, and it forms a part of the work of this book; however, we can already see that such a narrow inquiry into function by no means exhausts the tasks of evaluating and understanding the artifact model.

On the face of it, there are quite simple answers to most of the questions I've just raised. The evolutionary process bears deep similarities to the process of intelligent design. It is these deep similarities that explain and justify the appearance of the same vocabulary in both domains. Just as a designer chooses her materials to fashion an object to meet her problems, so nature selects traits to fashion an organism to meet problems laid down by the environment. Natural selection thereby plays a role analogous to intentional choice, and natural selection is what grounds various claims about function and design in the natural world. Since selection works only on organisms that reproduce themselves, it is selection that explains why artifact talk features in biology alone, and not in the physical sciences. Selection gives traits norms that should be met; hence selection can underpin normative function claims of the sort intended to ground projects to naturalize content in the philosophy of mind.

I will argue that such a picture is almost completely mistaken. There are deep similarities between the processes that go into the construction of organisms and artifacts; however, although these can help to explain why both types of objects enjoy a gradual accumulation of useful traits over time, it is a mistake to think that natural selection is a good analogue to the intentions of a designer. And it is the internal constitution of biological items, not the fact that selection acts only on biological items, that best explains the appearance of artifact talk in biology alone.

Much of the argument for these propositions turns on a demonstration of just what natural selection is. Natural selection is essentially a population-level, statistical phenomenon. Intentions, on the other hand, can have influence on individual entities. This does not mean that any other element of the evolutionary process yields a better analogue to intention that might instead be used to ground claims about function or design. We have a choice over just how we wish to tighten up function talk in biology, depending on what incongruities with our talk of artifacts we are prepared to tolerate. I suggest that function claims in biology are best understood quite simply as claims about contributions to fitness. However, whatever option we choose, the failure of biological processes to yield a function concept that closely matches the connotations of artifact functions puts limits on the burdens such a concept can bear.

The work of this book also has an impact on debates about creationism. What falls out from its treatment of artifact talk in biology, in terms of the constitution of organic nature and the processes of survival and reproduction, is an explanation of the use of language admittedly laden with connotations of intelligent design. No intelligent designer is needed to make sense of artifact talk.

1.2 Why Is Teleology So Boring?

I will give a map of the structure of this book toward the end of the chapter. First, I should say a little about my choice of topic. There seems to be a feeling among many of the most prominent philosophers of biology that the problem of teleology is a boring one, either because it has already been solved, or because there is no real problem beyond being clear about what one intends when one speaks of "function"; or, because the debate is fruitless, consisting for the main part in the exchange of intuitions about whether one would use the word "function" in certain artificial imaginary scenarios.

Let me give three examples. First, Michael Ruse, in the paragraph that forms one of the epigraphs to this book, tells us that the problem of teleology "is worked out. Natural selection produces designlike objects and so function talk is appropriate" (Ruse 1996, p. 284). The idea that there is little left to say on the subject is supported by the fact that teleology is one of the very few topics in philosophy where there is anything resembling a consensus. Almost all contributions to the functions debate over the past twenty years have consisted in refinements of Wright's (1973) etiological analysis. Examples of such approaches include papers by Neander (1991a,b), Griffiths (1993), Kitcher (1993), and Godfrey-Smith (1993, 1994) to name just a few. Two recent collections—Buller (1999) and Allen, Bekoff, and Lauder (1998)—are dominated by etiological analyses. This said, the most recent work on the topic (Ariew, Cummins, and Perlman 2002; Davies 2001; McLaughlin 2001) shows signs that some are moving away from the etiological consensus.

The basic innovation of recent etiological accounts has been to supplement Wright's analysis with an explicit reference to natural selection, and recent papers tend to argue only over just what the appeal to selection should look like. So while Wright's analysis tells us that the function of some item is what it does to explain why it is there, newer etiological analyses tell us, basically, that a biological item's function is what tokens of that type did in the recent past that caused them to be selected.

For Ruse it seems that the functions question is a significant one, but it has become boring because it has been answered successfully. It is the nontrivial fact that natural selection produces designlike objects that means that function talk is appropriate. Had selection not had this character, function talk would have been a mistake. A comment by Elliott Sober hints at a second type of complaint: "If function is understood to mean adaptation, then it is clear enough what the concept means. If a scientist or a philosopher uses the concept of function in some other way, we should demand that the concept be clarified" (Sober 1993, p. 86). Sober's apparent fatigue is, like Ruse's, partly a result of the thought that the problem has been solved—after all, most philosophers and biologists seem to agree that the analyses of function and adaptation should match—but it also expresses some puzzlement about why we should think there is a serious philosophical problem of functions at all. We need only be clear in saying what we mean by "function" in some context, and that is that. Sober's problem, then, seems quite different to Ruse's. For Ruse, the problem of teleology seems to be the substantive one of vindicating a potentially illegitimate vocabulary. For Sober, it seems to be one of giving clarity to words that are ambiguous.

Finally, David Hull (1998) characterizes the debate somewhat differently again. Hull is bored because he thinks of the literature on functions as a form of *conceptual analysis*—a project which he characterizes as the search for the meaning of some phrase like "S knows that P." One philosopher proposes a set of necessary and sufficient conditions to capture the use of the phrase, and other philosophers respond by concocting more or less elaborate scenarios in which the analysis fails to match with their intuitions regarding whether the word should really be used in that scenario or not. The analyst responds by modifying the analysis, or by asserting, like Nissen (1997, p. 215), that the other philosopher's intuitions "are just wrong."

The functions debate certainly has been conducted in this way by some protagonists. One hears what Bigelow and Pargetter (1987) memorably call "the dull thud of conflicting intuitions" in the following passages from Wright and Kitcher, respectively:

> If a small nut were to work itself loose and fall under the valve adjustment screw in such a way as to adjust properly a poorly adjusted valve, it would make an accidental contribution to the smooth running of that engine. We would never call the maintenance of proper valve adjustment the *function* of the nut. (Wright 1973, p. 63)

> Unbeknownst to you, there is a connection that has to be made between two parts if the whole machine is to do its intended job. Luckily, as you were working, you dropped a small screw into the incomplete machine, and it lodged between the two pieces, setting up the required connection. I claim that the screw has a function, the function of making the connection. But its having that function cannot be grounded in your explicit intention that it do that, for you have no intentions with respect to the screw. (Kitcher 1993, p. 380)

These are not peripheral to the philosophers' accounts; the intuitions they express dictate how their theories of function are formulated.

Perhaps some forms of conceptual analysis are legitimate. Thus, on some views of the meaning of scientific terms, we see meaning as deriving from the roles of those terms in the theories in which they feature. Now on this view, to say what the role of terms like "function" is in biology is also to give an account of the meaning of those terms in biology. In this sense it is a conceptual analysis. And the intuitions about use, of those well versed in the theory in question—the intuitions of biologists and well-informed philosophers of biology—could be essential to recovering the role of the term in the theory, and hence its meaning in this sense. Still, these will be intuitions about biological cases; it is hard to see what role there is for intuitions about screws in machines in uncovering the meaning of the *biological* function concept.

So for Hull, the source of frustration with the debate as a whole is different again. Here it seems what is at stake is neither the vindication of a problematic vocabulary, nor the attainment of conceptual perspicacity, but instead the provision of an account of what some concept really means. Comparison with Sober's implicit project will make this clear. For Sober, the project of giving a meaning to the word "function" is achieved just so long as a clear definition is given suitable to biologists' purposes. Giving a clear meaning to a term is not the same as saying what the existing meaning of a term is. At least that is so on an account of meaning according to which we uncover meaning by trying to match actual, rather than recommended, biological use. That is why the two projects respond in different ways to examples of actual use and imaginary scenarios. If our business is giving clear definitions, then considerable tension with both actual use and intuitions concerning imaginary cases can be expected and tolerated. That is true even if we are concentrating on a specifically biological function concept and looking exclusively at biological usage. If, on the other hand, we are trying to outline what the word's existing meaning is, then we should try to match intuitions—and certainly actual use—far more closely. Hull's problem is that the methods available for carrying out this project seem weak. What are we to do when intuitions conflict? Whose intuitions should we respect? Isn't this form of conceptual analysis an *empirical* project?

If philosophers can get frustrated by the functions debate in such different ways, then it suggests that they have quite different conceptions of what the goal of an account of functions is, and what the proper methods are for attaining it. Is it simply a question of bringing clarity to biology? Should we also ask what biologists in fact mean by their terms? Is there any more substantive issue at stake about the nature of design in the organic world?

The project of saying what biologists mean by their terms leads us into thickets that we can happily avoid in our goals of understanding how teleological approaches in biology work, what risks they carry, what forms of teleological content can be grounded by biological processes, and why teleological approaches are found only in biological contexts. These questions elude the complaints of Sober, Hull, and Ruse, for they are substantive, they do not require the idle exchange of intuitions to be answered, and they have not been solved already. But we might now fear that in giving up on the project of exposing existing meaning as too difficult, or subject to idle comparison of intuitions, we are then pushed toward saying our project is merely one of stipulation or construction of meaning suitable for some purpose. Sober suggests that the debate over functions is merely one of what concept is most appropriate for biological use. Millikan (1984, 1989c), notoriously, considers her analyses of function to be exercises in stipulation to be measured by the work they do in philosophy of mind. In neither case does the project seem like an interesting philosophical one for what it says about biology itself.

In fact, so long as our project is understood broadly enough, there is a way of approaching the phenomena of artifact talk in biology that reduces to neither a dull exercise in the comparison of intuitions nor a dull exercise in stipulation. In the next section I give an imaginary example that helps us see how our questions should be tackled, and also why we can be silent on the question of meaning.

1.3 Meaning, Metaphor, and Methodology

Our problem is to understand why biology makes use of a vocabulary that seems ill suited to it, to understand whether that vocabulary is genuinely helpful, and to understand why biology, and not the physical sciences, tends to make use of that vocabulary. I would like to use a parable to explain how we should go about answering these questions.

For many centuries scholars thought that all of nature was invested with spirits who controlled the movements of rocks, trees, clouds, and so forth. They would speak of rocks, trees, and clouds in human terms, reflecting the intentions of the agents who were thought to reside within them. So trees would strive to attain the sunniest spots in the forest, rocks would race each other downhill in landslides, and clouds would chase each other across the sky. With the development of physical theory, most of the sciences abandoned this animist paradigm. Now geologists would no longer talk of rocks racing each other down hills, only of some falling faster than others. And botanists gave up speaking of trees striving to attain the light, preferring instead to understand their motions as the result of mechanical tropisms. Only the meteorologists persisted with the old vocabulary. Still they would talk of clouds chasing each other across the skies. And the models they produced, framed in that vocabulary, had great predictive success. Thinking of clouds as chasing each other allowed them to predict successfully that the faster chasers would succeed in their goal of catching and swallowing the slower clouds. The chasing paradigm in meteorol-

ogy seemed to work perfectly well as a model for predicting how cloud positions and cloud conformations would change in the skies.

Philosophers of science found the meteorologists' success puzzling. How was it that they were able to persist with a vocabulary that had been discredited by the rest of science? Could it be that clouds alone really did harbour some kind of intentional or pseudointentional states that explained the survival of the animist paradigm in meteorology alone? And what should they say about the small group of renegade meteorologists who argued that the animist paradigm was misleading, that meteorology should grow up, that it should go the way of geology and begin speaking of clouds in the sterile ways the enlightened geologists speak of rocks?

It will be helpful to keep this parable in mind throughout this book, for it makes clear how philosophers stumble into analytical dead-ends in thinking about function and design that they would recognize quite clearly in the case of our meteorologists. How should our imaginary philosophers of science proceed?

If our goal is to explain how meteorologists have been able to continue to use the language of "chasing" with predictive success, then we need to look to similarities between the motions of clouds and the motions of people to explain that success. It is because cloud positions are covariant in ways that resemble the positions of people who chase each other, that "chasing talk" remains in meteorology.

A philosopher who maps these similarities does most of the work in explaining the persistence of chasing talk in meteorology. Yet he has not provided any kind of analysis of "chasing talk." He has not given us some short formula of the form "Two clouds chase each other iff conditions C obtain." Suppose he looks at the relations of covariance between clouds and decides that meteorologists typically use the language of chasing when certain kinds of covariance relations apply. He then supplies an analysis of the concept "meteorological chasing," which gives these covariance relations as necessary and sufficient conditions. How should we understand this analysis? Does it tell us what the meteorologists mean by "chases"?

This is the kind of question that has led philosophers of biology into a dead-end. Once we know what kind of chasing concept meteorological processes are able to support, we can then compare that concept with the concept we apply to humans chasing each other. We will conclude that they differ in some respects, and that they are similar in others. Thus, we can answer the questions of how meteorology is able to make use of the chasing vocabulary, how the chasing vocabulary might mislead through associations with human chasing, and whether there are deep similarities or mere vague resemblances between the meteorological and human chasing concepts. We do not need to say what meteorologists mean in order to answer these questions. The answers we give are compatible with the thought that on the lips of meteorologists, the word "chases" has a strict technical meaning that should not be confused with the vernacular concept, or that it is a metaphor that happens to be useful, or that their talk is wholly mistaken and they really believe clouds to be invested with spirits. Trying to say which of these views about meaning is correct adds very little to our understanding of chasing talk in meteorology.

What is more, if our view of meaning is that "chases" is metaphorical, then the tightened analysis of chasing is of great value, since it gives the scientists a cleaned up concept that tells them what steps they need to go through to test claims about chasing, and so forth. If we think the concept is a technical one, then the analysis has value in a similar way; it helps tighten any looseness in discourse. So, whatever our view of meaning, the proposed analysis has value, and for similar reasons. The question of the status of the analysis—whether it cleans up actual meaning, whether it exposes actual meaning, or whether it creates a technical meaning to replace a metaphor—need not be answered. Note, however, that if all we do is give a stipulative analysis of "meteorological chasing" that is intended to be useful to meteorologists, philosophers, or whoever, then we fall short of answering many of the questions that interested us at the outset about what the relationship might be between this kind of chasing and normal human chasing, or how the chasing vocabulary in the meteorological realm might mislead through inappropriate connotations. That is why the stipulative project on its own is of limited value, unless supplemented by a contrastive exercise that ranges across the two domains.

Without some kind of comparison between artifact functions and biological functions we have no guarantee that biological functions have more than the most distant relationship to their artifact cousins. Compare: we notice that physicists use the word "charm" in connection with quantum particles, and decide to stipulate some meaning for what charm means there. It is clear that what we might call quantum charm has nothing whatsoever to do with romantic charm, and can support few of its connotations. What is more, unless we can show that biological function and artifact function are closely related, we cannot claim to have provided anything like a "naturalized" account of functions. In giving an account of the meaning of quantum charm we certainly do not give a naturalized account of romantic charm—this is so precisely because romantic charm and quantum charm have nothing to do with each other. To give an analysis of "biological function" in terms of wholly natural processes does not consist in a successful naturalization of function unless one can demonstrate that the new concept merits the name it bears.

The methodological stance that I have developed in the context of meteorology and physics goes for biology also. Whether we think that the meaning of "function" is technical or metaphorical, we need to examine the similarities between the processes that underlie the production of organisms and artifacts. This exercise tells us what kinds of function concepts biology can support, and how close they are to the function concepts we apply to artifacts. By looking at the roles teleological terms play in biological research and biological theorizing, we will also be able to construct a clear analysis of terms like "function" that will be beneficial to biologists. This is a valuable exercise whether we think the analysis exposes actual meaning or fashions a new meaning.

Note, finally, that my skepticism about the value of the exchange of intuitions, and about some forms of conceptual analysis, does not deny that thought experiments—even quite outlandish ones—may have value in understanding the significance of some set of concepts. Thought experiments can play a role in teasing out the similarities and differences between the kinds of concepts we apply to artifacts, and the kinds of concepts we can apply to organisms. Abstract thought experiments will play a role in my discussion of functions in chapters 5 and 6.

1.4 Metaphorical or Technical?

It is just as well that we do not need to say what artifact terms mean in the mouths of biologists, if only because this project would be so difficult to carry out conclusively. A decent case can be made, for example, for saying that when biologists use terms like "function," "design," and so forth, they mean exactly what the rest of us mean when we apply these terms to artifacts. This is what one might say who espouses a Davidsonian theory of metaphor, and who thinks that all artifact language in biology is metaphorical. On Davidson's (1978) theory of metaphor there is no distinct metaphorical meaning beyond the usual meaning of the words contained in the metaphorical sentence. Most metaphorical sentences, like "Cesare is a wily fox," are therefore false; however, they are used to draw attention to certain similarities—in this case, between Cesare and a fox. If those similarities themselves run deep enough, then they might explain why, for the most part, treating Cesare as though he were a fox might be a useful way to approach him. In biology also, we might explain the continued usefulness of metaphors of purpose, function, and design by reference to the deep similarities between the processes that go into the construction of organisms and artifacts. So this appeal to metaphor is one way in which Nissen's (1997) analysis of biological function statements, according to which biological functions are what some agent intends a trait to perform, might be able to make sense of the success of function talk within biology.

There are at least three prima facie reasons for thinking that much of the teleological language used in biology may be metaphorical, yet none of them is conclusive. First, as I have already remarked, there has been remarkably little change in biologists' use of teleological language over the past two centuries or more. In *The Blind Watchmaker,* for example, Dawkins (1986) demonstrates through both his choice of title and style of exposition that he regards Paley's *Natural Theology* (1802) with admiration, primarily for Paley's ability to expose the quality of design within nature.

Other natural theologians use language remarkably in tune with modern biology. The following passage from the fourth *Bridgwater Treatise* would not sound too unusual on the lips of a contemporary adaptationist:

> Shell fish have their covering for a double purpose: to keep them at the bottom of the sea, and to protect them when drifted by the tide against rocks. Animals of the molluscous division, which inhabit the deep sea, and float singly, or in groups, as the genus scalpa, have a leathern covering only: because they are not liable to the rough movements to which the others are subject, in the advancing and returning tides. (Bell 1837, p. 280)

Presumably, when organisms were considered to be artifacts made by God, function language had the same meaning regardless of whether one was talking about the function of a fork or a frog's leg. And function language continues to be used in the same way. It is a prima facie strength of the metaphor theory that it explains continuity of use in a simple way.

One who thinks that biological function instead has a quite technical meaning within the science, and who therefore argues against the metaphor theorist, can also explain resilience of use in spite of change of meaning through the mechanism of the *dead metaphor*. Dead metaphors, importantly, are not metaphors. When a metaphor dies, a word that was previously metaphorical loses its old meaning and acquires a new one. Wright (1976) gives numerous examples: when we speak of a jackknifed lorry, we do not speak metaphorically. It is testimony to the death of the metaphor that one can know what a jackknifed lorry is without knowing what a jackknife is. We could argue that terms like "function" are able to retain the same pattern of use in biology, because as biologists grow to realize that the nature of systems to which these terms are applied is different to what they had thought, they adjust the meaning of those terms to reflect that realization.

The second reason for thinking that teleological language might be metaphorical is that it is most often found in biologists' popular works. This might suggest that the primary function of teleological terms is to illustrate the makeup and history of organisms and parts of organisms for nonspecialist readers. The following two passages appear in *Nature*—a journal intended for a wide scientific audience—and in a work for popular consumption, respectively.

> If there are ways in which mutation can increase the probability of survival within cells without effect on organismal phenotype, then sequences whose only "function" is self-preservation will inevitably arise and be maintained by what we call "non-phenotypic selection." (Doolittle and Sapienza 1980, p. 601)

> Natural selection may build an organ "for" a specific function or group of functions. But the "purpose" need not fully specify the capacity of the organ. Objects designed for definite purposes can, as a result of their structural complexity, perform many other tasks as well.... (Gould 1980, p. 57)

These passages support the thought that any account which tries to give a tight analysis of what biologists mean when using terms like "function," "purpose," and "design" is misconceived. We could argue on the basis of these passages that such terms are not serious scientific terms at all. The words "function" and "purpose" are often placed in inverted commas. This suggests that such language is not intended to be understood literally, or that the users are suspicious of the propriety of their own terms.

There may be a good deal of truth in all this, but again, we do not need to drop the idea that "function," at least, is a technical term in biology. It is quite natural to assume that when writing for popular audiences, biologists would use words like "function" and especially "purpose" and "design" with caution precisely because they would not want them to be confused with the common language namesakes, and they would certainly not want their readers to think that they are committed in any way to the view that the organic world is the product of conscious design. Even if "function" is a respectable technical term within biology, one would not want a lay reader thinking one had the intentions of a creator in mind when speaking of the function of the panda's thumb. What is more, in the first passage the word "function" is used in scare quotes, in part because it reflects a nonstandard use in biology itself. The idea that selfish DNA has any function at all will sound odd to some biologists who are perfectly happy with the idea that other traits do have genuine functions.

The third reason for suspicion that function talk is metaphorical is, that although biologists may often speak informally of function and design, these terms are seen quite rarely in technical journals. One seldom finds straightforward claims about the functions of specific traits in such publications. Terms like "design" and "purpose" feature even less often. In a technical article by Kingsolver and Koehl (1985), often cited by philosophers in support of the claim that the functions of traits can change over time, the authors use the word "function" rarely, and they decline to make any explicit claims about the function of the insect wing. Instead they prefer to discuss the evolution of the wing in terms of its "adaptive value." Moreover, in those cases where the word "function" is used in this paper, it is most often found in locutions like "functions as," or "serves the function of." The authors thus distance themselves from paradigm statements of function of the form "The function of the wing is. . . ," and instead make claims like the following: "Elongation of the wings first evolved in small insects as a result of selection for thermoregulatory capacity, followed by an isometric increase—either gradual or abrupt—in body size, after which wings could function as aerodynamic structures" (Kingsolver and Koehl 1985, p. 503). Even here, the proponent of the technical term view can make a number of responses. First, as Allen and Bekoff (1995) note, the English language offers many ways of speaking of functions without using the word itself; technical papers often contain claims that are equivalent to function statements, such as "Parents remove white eggshells to protect their young" (Drickamer and Vesey 1992, p. 23). What is more, many research papers explicitly ask questions about function, even when the explicit answer given does not use the word. Allen and Bekoff cite a paper by Gordon et al. (1993) whose title is "What Is the Function of Encounter Patterns in Ant Colonies?" Gordon et al. make no explicit claim about what the function is; yet they do make a clear implicit function claim (p. 1099): "An ant that suddenly encounters alien ants may be in danger . . . the sudden increase in [antenna] contact rate, though short lived, may be sufficient to generate a defensive response to intruders." If function claims are not made explicitly, that is in part because of epistemic caution; in the paper by Gordon et al., the investigators simply are not confident enough to make firm claims. Yet it is clear that inquiry after function is central to many biological disciplines—most notably ethology and behavioral ecology. The question "What is the function of a trait or behavior?" is one of Tinbergen's (1963) famous "four whys?," and probably the one that behavioral ecologists, and more recently evolutionary psychologists, have been most strongly motivated to answer. A search through recent scientific journals yields a range of titles like *Functional Ecology, Cell Structure and Function,* and so forth, all suggesting that the concept of function has a central role even in the technical practice of biology.

In summary, both the metaphor account and the account of function as a technical concept can be made plausible. What is more, it is not clear to me what methods one would need to choose between them. Since they are both intended as accounts of what a particular group of people means by some term, the best methods for adjudication would seem to be empirical. We would need to undertake interviews with biologists of varying types, subject a range of journals to textual analysis, observe language use closely in the context of lectures, seminars, day-to-day fieldwork, and informal discussion around the laboratory. And there is no guarantee that the meanings attributed would be univocal. The picture that might well emerge is one of a set of terms with different meanings in the mouths of different biologists. For some they are metaphorical, for others they have some more-or-less technical meaning so that "function," like "altruism," is a biological concept with a rather elastic connection to its common language namesake. The difficulties in answering the question of meaning, and the fact that answering this question is unnecessary to our main project, mean that I will remain agnostic on exactly what terms like "function," "design," and so forth mean.

1.5 What Lies Ahead

As the preceding discussion makes clear, the first task for understanding the presence and limitations of the artifact model is an investigation of the nature of the processes that explain organic form. This task is undertaken in chapter 2, where the received view of selection is outlined, together with its supposed relation to the phenomenon of adaptation. I argue, first, that the view of evolutionary theory as a theory of forces needs to be handled with care, lest we lose sight of the fact that natural selection and drift must be understood as population-level statistical phenomena. Second, I show that natural selection should be distinguished from selective forces, and that these selective forces can explain the emergence of adaptation only when they range over suitably organized entities. The upshot is that development plays as much of a role in the explanation of adaptation as selection.

Chapter 3 introduces the artifact model, and begins an assessment of the use made of it by the adaptationist program. I outline the adaptationist framework that conceives selection pressures by analogy with design problems, and traits by analogy with the parts of artifacts that are designed to meet such problems. I argue that the most common criticisms leveled against adaptationists do not, in fact, threaten the artifact model in general, for they highlight methodological difficulties in explaining and predicting the form of artifacts themselves. That said, there are a number of crucial disanalogies between selection and intention—most obviously in terms of the population-level nature of one, and the individual-level nature of the other. These disanalogies mean that artifact thinking can lead us to ignore drift, and also to underestimate the functional interconnectedness of organic, as opposed to artificial, design. I also tackle the more practical problem of whether artifact thinking—especially in the guises of reverse-engineering and adaptive thinking—faces epistemic difficulties. Here I suggest that evolutionary psychologists, in particular, should not expect the strategy of predicting adaptive solutions to problems laid down in ancestral environments to be of much use in uncovering the workings of the mind.

Chapter 4 looks to more radical challenges to artifact thinking from what we might term *constructivist* and *structuralist* camps. The first group, of whom Lewontin is the archetype, argues that no sense can be made of the crucial concept of an adaptive problem to which solutions might be developed, with the result that the concept of adaptation should be dropped in favour of recognizing a dialectical, constructive relationship between organism and environment. In response, I construct a concept of an environmental problem that can serve the purposes of the adaptationist program, while taking Lewontin's legitimate concerns into account. The structuralist camp—exemplified by Bryan Goodwin, but with allies in David Wake and others—argues for an elimination of teleological styles of argument altogether in favor of mechanistic explanations of form alone. I show that while a structuralist research program that looks to the explanation of form independently of adaptation may have considerable value, it is unlikely to wholly supplant functional biology.

Chapters 5 and 6 look to more traditional problems in the philosophy of biology regarding the nature of function statements and functional explanations. It is important to distinguish sharply between two questions. First, What is the best analysis of function claims in biology? Second, What explains why biologists make function claims but physicists and chemists do not? The parable of the clouds shows how such questions can come apart. An analysis of "chasing" in terms of covariance tells us the best way of cleaning up meteorologists' chasing talk. Such an analysis won't do to explain why only meteorologists talk about chasing. That question might be best answered historically, or by reference to the fact that clouds look a little like creatures, or by reference to the usefulness of the approach.

In chapter 5 I argue that the best analysis of function statements in biology is simply to think of the function of a trait as the contribution that tokens of that trait make to fitness. The concept is nonhistorical. Some (even most) philosophers say that only a historical function concept that ties functions of traits to their selection history can make sense of the normative and explanatory connotations of function claims. To these people I reply by suggesting that selection does not, in fact, meet these connotations particularly well, that the nonhistorical concept meets them well enough, and that in any case, it is not clear that all of the connotations that have been thought to be marks of teleological function claims should really be accepted.

Chapter 6 addresses the comparative question of why we find artifact talk in biology but not in physical sciences like chemistry. One might think that it is because only biological items are subject to a special force—natural selection—that gives rise to purposive states. Here I argue, on the contrary, that selection is neither necessary nor sufficient for the appearance of artifact talk. Inorganic "sorting" processes—the kinds of processes that sort ions bonding to the surface of a metal catalyst, or nuts in muesli, or stones on a beach—might also give rise to such talk. What is more, in cases where selection does not act, and where we might encounter artifact talk all the same, one cannot argue that such talk is mere "as-if" function talk, whereas biological functions are more genuine, purposive features of organisms. That is so because sorting processes support the same connotations—the connotations typically alleged to be the marks of bona fide functions—as selection does. The result, then, is that the account of functions in this book should be regarded as deflationary regarding the normative status of biological functions.

The final chapter turns the organism/artifact analogy on its head to look at the prospects for an informative evolutionary theory of technology change. Most commentators in this debate think either that the evolutionary view is false, even obviously so, or failing that they believe the successful application of evolutionary theory to technology will revolutionize the way we think about design, or marketing, or history, or economics. Neither of these views seems right to me. Artifacts do evolve, yet only a very abstract version of evolutionary theory that declines to comment about the broad character of selection pressures and the nature of cultural inheritance systems can be made to fit. The price for this abstraction is a corresponding lack of explanatory and predictive power when we try to apply evolutionary models to specific technological changes. In spite of all this, I end by outlining some possible lines for future investigation for technological evolutionists, and I show why looking seriously at how selection models can explain intelligent design will give discomfort to those creationists who want to contrast explanations of natural design that look to selection with those that look to intelligence.

The principles that we need to investigate to show us how the design of artifacts should be explained take us right back to the first theme addressed in this book—the relation between adaptation, selection, and development. Developmental organization itself is instrumental in generating complex adaptation. Hence an inquiry into the general principles of development and heredity may yield insights for the study of both organisms and artifacts. Our central analogy will remain ripe for investigation even when the book is done. Let us begin, then, at the end, with adaptation and development.

2

Why Is an Eye?

2.1 Two Dogmas of Evolutionary Biology

There is a long tradition in biology—especially in British biology—of fascination with the many remarkable adaptations in the natural world. Here the natural theologian William Derham (1732, p. 189) notes the excellence of mouths: "Let us begin with the mouth. And this we find in every species of animals, nicely conformable to the use of such a part; neatly fixed and shaped for the catching of prey, for the gathering, or receiving food, for the formation of speech, and every other such like use."

Darwin was similarly, and famously, impressed by the adaptations of finch beaks, and no one disputes that the eye is a fine piece of work, wonderfully suited to its purpose. All who have been impressed by traits like these have asked broadly the same question as Darwin (1996, p. 51): "How have all those exquisite adaptations of one part of the organisation to another part, and to the conditions of life, and of one distinct organic being to another, been perfected?"

Modern evolutionary biology and natural theology have tended to agree that chance alone is not good enough as an explanation. The hypothesis that a fine eye, so useful for detecting predators, food, or potential mates, might have arisen whole and by chance is regularly compared to the thought that a Boeing 747 might be formed from a storm in a junkyard. Most modern biologists would, I think, agree with Paley's (1802, p. 43) rhetoric: "What does chance ever do for us? In the human body, for instance, chance, i.e., the operation of causes without design, may produce a wen, a wart, a mole, a pimple, but never an eye."

Still, natural theology and modern biology part company in their positive answers to Darwin's question. For the natural theologian, chance is replaced by divine design. For the modern biologist, it is natural selection that explains adaptation.

The object of this chapter is to clarify the nature of the processes underlying adaptation. Getting a clear understanding of the nature of evolutionary processes is an essential first step if we are to understand how those processes invite and permit artifact thinking. This chapter offers a clarification and criticism of two dogmas of evolutionary biology—the idea that selection is a force acting on populations, and the idea that selection explains adaptation. I distinguish selection from selective forces, and I show that there is a sense in which selective forces explain adaptation, yet they do not do so alone.

2.2 Is Natural Selection a Force?

What is selection? Many presentations of evolutionary biology, and of population genetics in particular, tell us that it is a force acting on populations that results in changes in gene frequencies. Because of selection, some allele increases its frequency within a population. It is quite obvious how selection, thought of in this way, can explain the increase in adaptedness of populations. In a population of mosquitoes where some are pesticide resistant and others are not, if that population is then exposed to a pesticide, pesticide resistance will increase in frequency when the mosquitoes that are not pesticide resistant die. This is not to say anything about whether, or how, selection explains the origin of the pesticide-resistant trait. It is this question about the origin of adaptation that interests me in this chapter. What does selection, understood as a force acting on populations to change gene frequencies, have to do with explaining adaptations? What does shuffling gene frequencies have to do with bringing eyes into existence?

In this section and the next, I argue that we need to handle the reading of evolution as a theory of forces with care. It is important to distinguish sharply between the forces that act on individual organisms in a population, thus causing changes in the composition of the population as a result, and the strictly population-level notion of selection and drift as evolutionary forces. In section 2.4, I move on to clarify the senses in which selection explains the origin of adaptation.

The view that selection is a force acting on populations is laid out explicitly by Sober (1984a), and has roots in the work of Fisher (1958) and Mayr (1982), among others. If selection is a force it is certainly an unusual one. In all standard presentations, selection is described as a process that requires heritable variation in fitness among members of a population. Where there is no variation, there is no selection. Returning briefly to our mosquitoes, selection can act on a population in which some have pesticide resistance and others do not, but not in populations where all have pesticide resistance or none do. So selection is a force that can be extinguished without changing anything about the environment in which organisms find themselves, and without changing the constitution of any particular organisms in a population, but simply by eliminating variants in the population until only one type exists.

This "two-ness" (a term I owe to Elliott Sober) of selection is not in itself objectionable. After all, gravity is a force that is always exerted between two individual objects. However, selection, unlike gravity, is not conceived as a force that is exerted between two individuals, or even between two groups of individuals. Selection is exerted by environments, one presumes, and it necessarily acts on tokens of at least two types.

We can also put some pressure on the idea of selection as a force by looking at how selection and drift are distinguished. In many presentations of evolutionary theory, selection is understood to act just in case a population of entities demonstrates heritable variation in fitness. Yet these conditions also allow the population to change through drift. Drift is typically understood to be a form of sampling error—a population in which the fittest variant fails to go to fixation is one that changes through drift. This makes it look as though drift and selection are not forces but outcomes—the question of whether drift or selection acts on a population is determined solely by looking at the outcome of a series of births, deaths, and reproductions. This is not merely a claim about how we best discover whether drift or selection acts; rather, on this view, what it is for drift to act is for a population to have changed in such a way that it departs from expectation, and what it is for selection to act is for a population to have changed in such a way that it accords with expectation.

Sober (pers. comm.) thinks that these considerations are compatible with a view of selection and drift as distinct forces, so long as we distinguish sharply between selection and drift conceived as *products* and selection and drift conceived as *processes*. We need to find conditions that count as selection acting, or drift acting, that are not mere summaries of the product of a series of births, deaths, and reproductions. It is at this point in discussing both selection and drift that we see why we must move to a strictly population-level, statistical conception of the two.

On Sober's view, as I understand it, the force of selection acts whenever a population shows heritable variation in fitness. The force of drift, on the other hand, acts in inverse proportion to the size of the population. Even when drift is very strong in this sense, the population on which it acts may not show any departure from the outcome that we expect on the basis of fitness. That can be shown with a coin-tossing analogy. The longer a series of tosses of a fair coin, the more likely the series is to reach a 1:1 ratio of heads to tails. Consider two sets of ten tosses. In the first set the coin lands heads up 9 times out of 10. In the second, the coin lands heads up 5 times out of 10. In terms of outcomes, the first sequence is more "drifty" than the second—it shows a greater departure from expectation. Yet Sober must say that the force of drift is equal in both cases. And Sober must also say that a third sequence of 100 tosses, in which the coin lands heads 99 times, is subject to less drift (understood as a force) than either series of 10, even though the outcome is much "driftier" than either short sequence.

This discussion helps to strengthen my claim that if we want to think of drift and selection as forces, we have to think of them as strange ones. Sober himself understands this, and in some ways I feel my coauthors and I could have done more to acknowledge this in Walsh et al. (2002). He remarks that "It is not controversial that two 'factors' influence whether the percentage of heads one gets on a run of independent tosses will fall in a given interval. These are (1) the coin's probability of landing heads and (2) the number of times the coin is tossed. But it would be bizarre to treat these two factors as constituting separate processes or forces" (1984a, p. 115). True enough, small populations are more likely than large populations to change in ways that depart from expectation based on fitness. And a population in which the difference in fitness between two types is large will change more quickly than a population in which the difference in fitness between two types is small. These differences are quantifiable, and they may yield opposing tendencies. In these respects, force talk is justified, although Sober also remarks that "nothing much hangs on this terminology" (p. 117). It is, however, important not to be misled by this kind of talk. Drift is no distinct force that acts on individuals. Take a series of ten coin tosses in which the coin lands heads eight times. In a sense we can say that the small size of the trial sequence makes the "force" of drift quite strong. This does not mean that there is some distinctive set of physical forces that acts on the coin in this series. The physical forces that act on the coin to determine its orientation are exactly the same in long and short series of trials, and in series of trials that depart from expectation and those that conform to expectation.

We have arrived at one important result. Discussions of drift often proceed as though drift is caused by things like lightning strikes, while selection is caused by things like predation. There are no grounds for this claim. The forces that explain the individual events in drifty series of births, deaths, and reproductions can be the same forces that explain selective series of births, deaths, and reproductions.

Walsh (2000) is especially keen to downplay the idea of selection as a force. As he points out, selective changes in trait frequencies or gene frequencies come about through the differing propensities of different types of individuals to reproduce: "Natural selection, it seems, is merely the consequence of an assemblage of causal processes taking place at the individual level. There is no need to invoke a distinct force operating over populations in order to explain the sort of changes in gene frequency thought to be explained by natural selection" (p. 139). Put this way, it seems almost too obvious that selection is not a force. It should at least make us look for an explanation for why the notion that selection is a force operating across populations has such appeal.

As we have already seen, population size and the strength of selection coefficients both make quantifiable differences to how a population is likely to change. It is quite natural then, to refer to these influences as forces. Also, our discussion makes it clear that the explanatory categories of selection and drift must be invoked at the level of *populations*. Even when the facts of which forces act on individuals are elaborated, some further explanation is needed for facts about sequences of births, deaths, and reproductions in the population as a whole. This can be made vivid by looking at another coin-tossing analogy. Suppose we agree that each landing of a fair coin can be wholly explained by looking to the microlevel forces acting on that coin. Even when we explain each individual instance of heads and tails by looking to these forces, we need a further explanation for why some types of sequence are more likely than others.

Why, for example, do we see sequences of tosses that show approximately fifty percent heads and fifty percent tails quite often, but long sequences that show ninety percent heads and ten percent tails are seen comparatively rarely? Such facts can either be written off as without explanation—there is nothing more to be said than that tosses happen frequently to follow this pattern—or, if we do explain them, we invoke chancy facts like the fairness of the coin and its propensity to land heads half of the time.

The coin-tossing example shows that we cannot eliminate the need for probabilistic explanations by claiming that individual tosses are wholly determined. A residue of facts about the regularity of certain kinds of sequence remains unexplained—regularities that chancy properties like the fairness of the coin can explain. Similarly, any claim that local ecological variables determine the survival and reproduction of each individual organism in a population does not show that evolutionary theory can dispense with chancy properties like fitnesses in explaining features of populations. Explanations in terms of selection and drift are just such population-level statistical explanations.

2.3 Selective Forces and the Force of Selection

It is important not to run together the physical forces that act on individual organisms and the population-level understanding of what it means to speak of the force of selection. There is a multiplicity of forces that act on individual members of a population, which may give rise to changes in the frequency of types in that population. These forces do not have the peculiarities of the force of selection. They can act on one type of individual alone, and they can act to produce changes that are identified with drift or changes that are identified with selection. A reason to be wary of adopting Sober's view of evolution as a theory of forces is that unless we are careful, it is easy to slide from "selection pressures" or "selective forces" thought of as a heterogeneous collection of forces that act on individuals, to selection thought of as an evolutionary force acting on populations.

Consider a simple example. Two variants of insect reside in a warm climate. One type is dark in color, the other light. As the sun shines on both variants, the darker insects tend to overheat, while the lighter ones maintain a healthier temperature. The darker insects expend greater energy in avoiding the sun, and miss out on mating opportunities as a result. In this case the sunshine is a "selective force" that acts across the population: it has a different effect on individuals of different types, so that the types reproduce at different rates. If only one variant were present, the same force would have the same effect on that type. (For simplicity's sake I am ignoring cases where the types interact with each other in such a way that the presence of the sun really would have a different effect if only one type were present.) With only one type present, the sun would either cause the lighter ones to survive at a healthy temperature, or the dark ones to overheat. The same force would act in the absence of variation, yet it would not give rise to selection, and in virtue of this we would be unlikely to speak of it as a selective force.

Once the category of selective forces is granted, we might think that we can identify the force of selection with selective forces. In fact, the identification of selective forces with selection yields awkward consequences for the relationship between selection and drift for reasons we have already seen. The very same individual-level forces that act to change the composition of populations in accordance with expectations based on fitness can also act to change the population in ways that deviate from expectations based on fitness. Hence, one who wants to identify selection with selective forces would have to say that selection sometimes causes drift.

The two levels of explanation—that of selection and selective forces—are related. In some cases the question of whether selection acts can simply be read off from the average fitnesses of the different types of organisms in the population, which, in turn, are determined by selective forces acting in different ways on the individuals bearing the traits in question. Yet selection can act even when the average fitnesses of different types are identical. Consider the following example taken from Charlesworth and Giesel (1972), in which the differential action of selective forces plays no role. When a population is expanding (when it is growing in terms of absolute numbers), the fitter type is that which has its offspring early in life, rather than late. When a population is contracting, the fitter type is that which has its offspring late in life, rather than early. The reason for this can be seen through an analogy with investment. Take two investors, one who saves money early in the financial year, another who saves an identical sum of money late. Will the early investor's money, expressed as a proportion of the total wealth shared by both investors (that is, the frequency of the early investor's money in the global fund) increase or decrease over time? The answer is that it depends on whether interest rates are positive or negative, which in turn is to say that it depends on whether the global fund is increasing or decreasing in absolute size. If rates are positive, the early investor's money increases its frequency in the "population"; if rates are negative, the late investor's money increases in frequency. Or, in the language of population genetics, where population growth is positive, there is selection for early reproductive investment.

Whether early or late reproduction is selected does not depend in this case on the differential action of forces on individuals. The forces acting on early and late reproducers do not cause the different types to have different numbers of offspring—not even in the long run. Indeed, the individual fitnesses of early and late reproducers—understood as the expected number of offspring of a representative individual of each type—are identical, as are the average fitnesses of organisms of each type. Yet one of these traits can be selected all the same, depending on facts about the population as a whole. The moral to draw from this is that there are cases where we can speak of selection acting on a population in virtue of different expectations for frequency changes of different types, yet selective forces act identically on the individuals of the different types. Selection and selective forces need to be sharply distinguished.

2.4 Does Selection Explain Adaptation?

Having gone some way to clearing up what selection is, we can move on to looking at how selection explains adaptation. In this section I will focus on explanations of adaptation in terms of selective forces. As we have seen, such forces can change population composition in favor of some trait type. So our question becomes: How do selective forces explain adaptation?

First, we need to distinguish the way that selection is thought to explain adaptation from the way a random search explains adaptation. Pharmaceutical companies sometimes use the technique of high-throughput screening of combinatorial libraries for the discovery of new drugs. Large collections of millions of molecules—called "libraries"—are generated more or less at random from combinations of basic constituents (hence "combinatorial libraries") and passed through a series of screens which test for desired functions. If the process works well, then a molecule with some important pharmacological function is thrown up by the screens. In a sense, we could say that selection explains, in this case, the adaptedness of the final molecules because the selective forces exerted by the screens ensure that only molecules with a good fit make it through.

There is an important difference between this kind of search and the search that selection is able to perform in the organic realm. In the case of a search through combinatorial libraries, the functional excellence of the resulting molecules is explained not by the gradual accumulation of good design, but by the vast size of the random library supplied to the screens. In a library so large we should expect functional molecules to exist somewhere. The screens then find those molecules. Natural selection, when it works on organismic lineages, is taken to explain adaptation in a different way. Nature does not test a vast number of variants against a selective environment until a mutant with a fully formed eye turns up. Indeed, typically biologists argue that this type of process could not explain the extraordinary wealth of adaptation in the living world. As we saw earlier, most modern biologists agree that "mere chance" does not suffice as an explanation for adaptive complexity. Instead, selection explains adaptation in virtue of its cumulative nature. This was Darwin's distinctive answer that made selection a plausible alternative to divine design in the explanation of complex adaptive traits: "Natural selection is daily and hourly scrutinising . . . every variation, even the slightest; rejecting that which is bad, preserving and adding up all that is good; silently and insensibly working, whenever and wherever the opportunity offers, at the improvement of each organic being in relation to its organic and inorganic conditions of life" (Darwin 1996, p. 70).

In both a random search and the case where selection is "cumulative," selective forces act in a broadly similar way. They favor organisms with certain propensities, so that individuals with those propensities will tend to survive the environment over others which do not. The pharmaceutical case is the same: the different propensities of molecules to survive various screens are able to explain how, if present in the set of tested molecules, a molecule with some functional effectiveness will be discovered. Yet in the pharmaceutical case there is no sense in which selective forces help to create these molecules. Selective forces only find good variants when they are already present.

So how does selection help with the creation of new variants? An appeal to the idea of a *selection pressure* does not give a satisfactory answer to this question. To say that there is a selection pressure for some function F does not mean that one type is outreproducing another in virtue of some effect F; instead, it tends to mean that as organisms get better at the effect F, they will continue to increase their fitness. There can be a selection pressure in this sense even when there is no variation (and no selection), for the environment may be such that were an organism to arise with certain properties, then it would tend to increase its representation. Still, this kind of dispositional fact about an environment does not, on the face of it, make any particular type more likely to be created. Again, the appeal to selection seems to explain only how variants, once created, are found.

We arrive again at our original question. What is the relationship between selective forces thought of as causes of the differential reproduction of certain variants already present in a population, and selection thought of as a cumulative process whereby new adaptive designs are efficiently created? How does the first conception of selection relate to the second? We have seen that selective forces can increase the frequencies of genes. So the idea must be that by increasing the frequencies of genotypes (or phenotypic traits), selective forces thereby make certain further combinations of genes more likely. That is why selection is credited with creative power. Ayala hints at what selection has to do to be creative: "Natural selection has been compared to a sieve which retains the rarely arising useful mutations and lets go the frequently arising harmful mutants. Natural selection acts in this way, but it is much more than a purely negative process, for it is able to generate novelty by increasing the probability of otherwise extremely improbable genetic combinations. Natural selection is creative in a way" (1970, p. 5). A suggestion of exactly how selection achieves this is provided by Karen Neander: "Selection does more than merely distribute genotypes and phenotypes . . . : by distributing existing genotypes and phenotypes it plays a crucial causal role in determining which new genotypes and phenotypes arise" (1995b, p. 585).

Here, roughly, is what Neander has in mind (see 1995a, p. 77). Suppose we have three "genetic plans"—P1, P2, and P3. P1 codes for a proto-eye, P2 for a slightly better eye and P3 for an eye that is better still. Now take a population that is entirely composed of P1 individuals, with the exception of one which has the P2 mutation. P2 is fitter and so spreads through the population. Neander's idea is this—as the frequency of P2 increases, so the chances of a P3 mutation increase also. So, a selective force that increases the frequency of the genetic plan P2 thereby increases the chances of some P3 mutation arising in some member of the population. This is how selection, by distributing genotypes and phenotypes, also explains the emergence of complex adaptations.

Of course Neander is not making the mistake of saying that selection directs mutation in individuals—selection simply increases the sample size over which some individual might get the P3 mutation by chance. If selection can increase the number of P2 individuals, then selection can also increase the number of chances P3 has of arising in those individuals. Selection does direct mutation in a population, in the sense that it makes mutations more likely to occur in individuals of the selected type than in the type selected against.

One preliminary should be noted. There is an ambiguity in saying that selection increases the chances of some P3 mutation arising. Selection of or for a trait or gene indicates that the trait or gene will tend to increase its frequency. Now this may, or may not also, involve a tendency for the trait or gene to increase in absolute numbers. So P2 could be selected in spite of the fact that its absolute numbers decrease, for example, when other variants are in more rapid decline. Equally, the absolute numbers of a trait might increase when it is being selected against (for example, in a case where there is no scarcity of resources). Finally, the absolute numbers of a trait might increase when there is no selection because there is no variation.

These observations have two important consequences for the supposed role of selection in explaining adaptation. If selection is taken to explain the emergence of some P3 mutant in the sense that it increases the absolute number of P2 individuals, then we must acknowledge that selection only sometimes explains adaptation, and that selection is not required for the explanation of adaptation. Selection can increase the frequency of some trait even when the absolute number of the trait decreases; here we should say that selection, and selective forces, decrease the chances of adaptation. And the absolute numbers of some trait can increase even when there is no selection, or when the trait is being selected against. Here we should say that the reproductive success conferred by earlier traits increases the chances of, and thus explains, adaptation, irrespective of selection. In this sense, the role of selective forces in explaining adaptation is merely a special case of the role those same forces can have when they are not labeled selective, because there is no variation.

In spite of all this, there is a different sense in which a selective force that increases the frequency of a genetic plan explains adaptation, even when absolute numbers are decreasing. The idea is, I take it, that selection makes a search more efficient. If there are two types, P1 and P2, and if P2 is fitter than P1, then, because there will be more P2 individuals than P1 individuals, there are more chances for beneficial mutations to arise in the P2 individuals than in the P1 individuals. Selective forces ensure that in a population, the already fitter variants are searched more thoroughly for further adaptive mutations than the less fit variants.

This is how selection makes the creation of adaptation more likely than a random search and hence how selection is able to be "creative," as Ayala puts it. Selection makes adaptation more likely than a random search, because if one is limited to a finite number of trials, one is better off trialing the already fitter variants in the hope of finding ones that are fitter still, than picking variants at random.

We can now see more clearly the extra assumptions that Neander needs if selection is to increase the chances of P3 arising rather than decrease them. The idea behind Neander's claim is that selective forces ensure that a population is always looking for new mutations with the greatest efficiency. The P3 mutation is more likely to arise in a population of mainly P2 individuals than in a population of mainly P1 individuals. But this claim in turn relies on the idea of cumulativity itself—that it is easier to make an eye from something that is already rather like an eye than it is from something that is nothing like an eye. This need not be the case. Suppose, for example, that the P3 genetic plan is more similar to the P1 plan than it is to the P2 plan. Then, by increasing the proportion of P2 individuals in the population, selection decreases the chances of the P3 mutation arising.

It is only when genetic plans are related in such a way that plans for fitter phenotypes are easier to access by mutation from slightly-less-fit plans than from much-less-fit plans, that the strategy of searching the fittest available phenotypes in a population is the right one for finding phenotypes that are fitter still. If selection is ever to result in adaptive complexity, we must also assume not only that P1, P2, and P3 are plans for progressively better eyes, but that these plans really do make the whole organism fitter at each step. Suppose P3 is easier to access from P2 than from P1. Even so, if P2 yields a more visually acute eye than P1, but that eye also drains too much energy, or it interferes with other sensory organs, or it reflects light and attracts predators, then P2 won't be selected after all. Here selection decreases the chances of adaptation in the sense that selection for some property—predator avoidance, say—tends to stop selection for visual acuity from bringing the P2 plan to fixation, hence, it stops selection for visual acuity from bringing the P3 plan into existence. When an organism is troubled by these kinds of conflicting trade-offs, the complex interactions of selection pressures can stop any one of them from bringing any kind of complex adaptation into existence. This is a way of stating the familiar requirement that for complex adaptations to evolve by selection there must be a smooth series of variants, each of which is better than the last—not merely in terms of how it performs its local function, but in terms of how it interacts with the whole organism.

It is an entirely contingent matter whether the genetic plans for increasingly better eyes, P1 to P3, really do make the organism as a whole fitter, and whether the better plan is always more easily accessible from the next best plan than from any worse plan. Unless these assumptions are met, then selection, by changing frequencies of traits, can make the emergence of complex adaptations not more likely, but less likely.

These assumptions themselves are assumptions about the nature of the individuals undergoing selection. So, we can summarize: only when selective forces act on individuals with certain characteristic properties can these forces produce complex adaptations. Alternatively, only individuals which vary in their propensities to survive and reproduce and which also have further characteristic properties will accrue complex adaptations. Now we do not need to conclude that selective forces never explain adaptation. But we should conclude first, that selection alone does not explain adaptation, and second, that selection only explains adaptation in rather tightly circumscribed contexts. Selection is not by itself "cumulative"—the outcomes of selection processes have this feature only when the items undergoing selection have certain properties.

2.5 Universal Darwinism

The cumulativeness of the evolutionary process is sometimes presented as though it were an automatic consequence of the selection of fitter variants, where heritability explains how, once discovered, "good tricks" (Dennett 1995) will be preserved and improved upon. Dawkins (1986) and Simon (1996) both use similar examples to illustrate how selection improves on random search techniques for the discovery of solutions to complex problems. Simon's example (1996, p. 194) is of a combination lock. If one is trying to crack a combination safe lock with ten wheels, and one proceeds by spinning all the wheels at once, then one is very unlikely to find the correct combination. If one instead spins each wheel in turn, and stops when the wheel shows the right number, then the lock can be cracked far more quickly.

It seems quite likely that those universal Darwinists who stress the existence of selection processes throughout nature (in culture, in the brain, in the generation of universes), and who have tried to generate general accounts of selection to capture all of these diverse processes, have been motivated to undertake this work because they have believed that selection processes are special by virtue of their abilities to produce complex adaptation (e.g., Czicko 1995; Dennett 1995). It is true, as Darden and Cain point out in their general analysis of "selection type theories," that "selection theories solve adaptation problems by specifying a process through which one thing comes to be adapted to another thing" (1989, p. 106). Yet it is important to recognize that the existence of a selection process does not guarantee the emergence of complex adaptation. It constitutes part of the explanation, but not all. This is a fact that Kauffman (1993, 1995) makes much of, but which Lewontin (1978) anticipates in his requirements of *continuity* and *quasi-independence*.

Kauffman uses statistical models to show that systems with many elements will be hard to improve through a selective process if each of the elements is affected by all of the others. In such circumstances the chances are high that if any one element is altered, adverse interactions with other elements will lead to an overall reduction in function. Hence gradual improvement of any one trait is likely to be vetoed by the corresponding decrease in performance, or outright disruption, of many other functions of other traits. So in systems that are highly integrated, cumulative evolution is unlikely to occur.

Lewontin's condition of quasi-independence expresses the same observation. The condition of continuity adds the requirement that small mutations to the phenotype should not lead to grossly different ecological relations, hence to wildly varying overall fitness. The requirement of quasi-independence demands that development enables the building of specialized parts of the phenotype; the continuity requirement demands that phenotypic variation map smoothly onto variation in fitness via smooth changes in ecological relations.

I do not want to go into Kauffman's work in any detail. The one point that is worth making is to remind the reader that Kauffman's preferred mode of explaining the conditions for evolvability in terms of the structure of *fitness landscapes* apt for cumulative evolution is equivalent to a claim about how individual organisms in the evolving population must be structured. A fitness landscape, originally devised by Sewall Wright (1932), is a device for representing the relative fitnesses of genotypes. It is a kind of map, with distances between genotypes corresponding to their allelic differences. Genotypes that are near to each other on the map are only a few mutational steps away; genotypes that are at larger distances from each other require more mutational steps to move from one to the other. Finally, genotypes are assigned fitnesses, measured by the altitude of the landscape. The result is a surface that shows the fittest genotypes on adaptive peaks, with the unfit genotypes in troughs or valleys. The facts that determine what type of fitness landscape organisms inhabit, hence whether they can undergo cumulative evolution by natural selection, are often facts about the systemic organization of those organisms themselves. If a population is on what Kauffman calls a "rugged" landscape—one where neighboring genotypes have wildly differing fitnesses—then this can be because organisms in the population are structured in such a way that the alteration of any one part will tend to disrupt the functioning of all other parts.

One might think that selection will tend to find the right kind of internal organization required for adaptability or evolvability. After all, internal structures can evolve just as well as relations to the external environment. So given enough time, a selection process will ensure the emergence of complex adaptation, by first ensuring the emergence of the right kind of internal organization for evolvability. I have two comments in response. First, showing that developmental organization is itself amenable to selective explanation does not show that development plays no role in explaining adaptability. Second, while selection may explain how a system with the right developmental organization goes to fixation once it first appears, it is hard to see how selection could shape systems to be suitable for cumulative evolution. For this selective "shaping" to work, systems already need to be apt for cumulative evolution. This does not mean that we need to invoke intelligent design, or a brute miracle, to explain the original emergence of evolvability. We might instead try to show that evolvability is not such a surprising property for systems to possess after all—indeed this seems to be Kauffman's response to the problem (1995, p. 188). In any case, we retain an important role for nonselective explanation in our attempts to understand adaptation, and we thereby remind the universal Darwinist not to assume that securing the existence in some domain of a selection process thereby discharges the task of explaining adaptability in that domain.

Let me conclude this chapter by summarizing its main findings. First, we need to distinguish between selective forces, understood as physical and other forces that can alter the makeup of populations through their action on individuals, and the force of selection, which must be understood as a wholly population-level, statistical phenomenon. There are cases where the force of selection cannot be reduced to talk of the differential action of selective forces on different kinds of individuals.

Second, although selective forces explain adaptation, they do not do so alone. The right kind of developmental organization is also needed if systems are to be apt for cumulative evolution. This kind of organization is partly described through the conditions of quasi-independence and continuity. Now that we understand better the processes of organic evolution, we can look at how the artifact model of evolution—the model that treats organisms as though they were artifacts—is laid onto them.

3

Adaptationism and Engineering

3.1 Introducing the Artifact Model

In this chapter I elaborate in detail what I call the artifact model of evolution. This, briefly, is the approach to the organic world that treats it as though it were designed, by speaking of environmental problems, organismic solutions, the purposes of traits, and the design of adaptations. The question of the merit of these ways of thinking is often discussed in the context of evaluations of adaptationism, so I should begin with a few words about what adaptationism is.

Peter Godfrey-Smith (2001) recognizes three types of adaptationism, and in a different paper (Lewens forthcoming-b), I find seven types. It is good enough for our purposes in this chapter to distinguish two broad varieties—hypothetical adaptationism and heuristic adaptationism. For some, the debate over adaptationism is an empirical question about the power of natural selection. Under this interpretation, adaptationism becomes a hypothesis, and one that can have various strengths. A familiar caricature of the adaptationist hypothesis (more accurately, a caricature of anti-adaptationists' construals of the adaptationist hypothesis) says that all organisms develop the best conceivable traits in response to problems posed by the environment, so that zebras should evolve machine guns to fend off predators. A milder adaptationist hypothesis, and one which is not obviously false, is expressed in Orzack and Sober's (1994) and Sober's (1998) conception of adaptationism as the thesis that "Natural selection has been the only important cause of most of the phenotypic traits found in most species" (Sober 1998, p. 72). Others (e.g., Resnik 1997) think of adaptationism as a heuristic—a recipe for generating hypotheses and conducting research. Considered in this mode, adaptationism has no truth value, but should instead be judged by its fruitfulness.

My goal for this chapter is not to question the truth or falsity of hypothetical adaptationism, nor to argue over the heuristic benefits of proposing specific adaptationist hypotheses. Rather, I want to examine a set of investigative tools that often form part of the adaptationist's kit. Adaptationists typically recommend using a suite of techniques that treat organisms as more or less well-designed solutions to environmental problems. Following Griffiths (1996), we can distinguish two general forms of artifact thinking: *reverse-engineering* seeks to infer both the problems posed by an organism's environment and the constraints on what solutions could be adopted to those problems from data regarding observed organismic traits. *Adaptive thinking* reverses the direction of inference and seeks to use knowledge of adaptive problems faced by an organism to predict likely solutions that will have emerged to meet those problems. The bulk of this chapter consists in an evaluation of these modes of thinking, and the general merits of the artifact model. I also consider the technique of optimality modeling. Optimality models are a particularly formal version of the general technique of reverse-engineering, and they have received especially strong criticism for their alleged Panglossianism.

This artifact model of adaptationism has received its strongest formulation and most vigorous support in several works by Dennett (1983, 1988, 1990, 1995). Dennett in fact makes two claims; first that it is fruitful to investigate organisms as though they were artifacts, second that it is necessary to do so. Not only is adaptationism useful, says Dennett, it is the only stance for any legitimate biologist to adopt. On the usefulness of the artifact model Dennett writes:

> Instead of trying to figure out what God intended, we try to figure out what reason, if any, "Mother Nature"—the process of evolution by natural selection itself—"discerned" or "discriminated" for doing things one way rather than another. (Dennett 1995, p. 213)

On the necessity of the position he claims:

> Adaptationist reasoning is not optional; it is the heart and soul of evolutionary biology. Although it may be supplemented, and its flaws repaired, to think of displacing it from its central position in biology is to imagine not just the downfall of Darwinism but the collapse of modern biochemistry and all the life sciences and medicine. (Ibid., p. 238)

One might assume that Dennett's account of evolutionary biology would have few adherents. Those sober scientists who count themselves adaptationists would surely consider the introduction of the vocabulary of purpose and design a step back towards natural theology. In fact, artifact thinking of one form or another has a strong pedigree within the adaptationist community. George Williams (1992, p. 190) recommends that biologists read Paley's *Natural Theology* for advice on how to recognize the products of selection. Ernst Mayr comments that "The heuristic value of teleological *Fragestellung* makes it a powerful tool in biological analysis, from the study of the structural configuration of macromolecules up to the study of cooperative behavior in social systems" (1974, p. 114). Pinker (1997) enthusiastically takes on the project of "reverse-engineering the human mind," and the importance of optimality models for the interpretation of selective forces and constraints on adaptation is proclaimed by Stephens and Krebs (1986), among many others.

The artifact model is, however, controversial. Anti-adaptationists such as Ghiselin eschew teleological thinking in general, and also the specifics of models that treat organisms as designed. So, he tells us that "Panglossianism is bad because it asks the wrong question, namely, What is good? . . . The alternative is to reject such teleology altogether. Instead of asking, What is good? We ask, What has happened? The new question does everything we could expect the old one to do and a lot more besides" (1983, p. 363).

Optimality modeling has been criticized from many quarters (e.g., Lewontin 1984; Ollason 1987), and Griffiths (1996) has argued that the techniques of reverse-engineering and adaptive thinking—so prominent in Dennett's exposition of adaptationism—are liable to mislead. Finally Lauder (1996) has argued that the use of optimal design as a criterion for the action of selection is a version of Paley's design argument that inherits many of the problems of its ancestor.

The partial defense I give of the artifact model in this chapter will be cold comfort to many adaptationists. I do not try to justify the unreflective use of optimality modeling, reverse-engineering or any other technique that might be suggested by the conception of organisms as artifacts. We will see that all of these techniques can mislead if applied crudely; however, this fact does not undermine the artifact model precisely because the same techniques often mislead us when they are applied unreflectively to artifacts.

Since many traditional criticisms of adaptationism can be phrased within the vocabulary of the artifact model, we will see that an adaptationism that recommends only the strategy of "treating organisms as though they were artifacts" (Dennett 1995) is an extremely weak position, and one that encompasses much of what anti-adaptationists believe. Many of the criticisms voiced by anti-adaptationists are best understood not as complaints against the artifact model itself, nor as claims about the relative unimportance of selection, but as expressions of skepticism about our abilities to uncover the evolutionary past. At least some of the conflict between those who call themselves adaptationists, and those who call themselves anti-adaptationists is, I suspect, a fight over nothing.

3.2 Weak Reverse-engineering

The artifact model advocates an investigation of nature using the assumption that evolution follows a designlike process that can be understood and predicted in the same ways that we understand and predict the processes of intentional design. Central to the application of the artifact model is the enterprise of reverse-engineering. What is this technique? Dennett offers the following description:

> When Raytheon wants to make an electronic widget to compete with General Electric's widget, they buy several of GE's widgets and proceed to analyse them: that's reverse engineering. They run them, benchmark them, X-ray them, take them apart, and subject every part to interpretive analysis: Why did GE make these wires so heavy? What are these extra ROM registers for? Is this a double layer of insulation, and, if so, why did they bother with it? Notice that the reigning assumption is that all of these "why" questions have answers. Everything has a *raison d'être;* GE did nothing in vain. (1995, p. 212)

What exactly do we learn by going through this process? I begin in this section by pointing out that there is a weak form of reverse-engineering that is independent of traditional adaptationist concerns with the prediction of adaptive change and the retrodiction of selective environments.

Let us begin by thinking about artifacts. What does Raytheon learn when they reverse-engineer a GE widget? If their engineers proceed under the assumption that every part has a raison d'être, the result is that any contribution that any part makes to the overall running of the gadget is likely to be revealed and can be copied when Raytheon tries to make its own gadget. This kind of interpretative endeavor need not depend on identifying the true problems addressed by GE engineers if it is to succeed in demonstrating how different parts contribute to the overall performance of the gadget. One's suppositions about the design problems faced by the engineers could even be false. Some of the contributions made by the parts of the gadget may not have featured in Raytheon's conception of the design problems it addresses. Yet proceeding as though there were some intention behind every effect can be useful nonetheless in illuminating surprising contributions to overall function.

It is quite easy to see how a similar heuristic might be used for the investigation of organisms. Just as we assume that all parts of GE's gadget contribute to its smooth running, so we might assume that all the parts of an organism contribute to some complex capacity—most usually, but not exclusively, survival and reproduction. This form of reverse-engineering might usefully be described as Kantian (see Kant 1952). One proceeds under a principle that organisms have been perfectly designed for the maximization of fitness. This principle is "merely regulative"; it is a principle that organizes and directs our inquiry to teasing out the details of how some system is orchestrated, but whose truth is not essential to the inquiry itself. This kind of endeavor is heuristically modest, since it proceeds simply by fixing on some capacity to be explained, and assuming that many diverse parts may have a role in that capacity. What this means is that just as we might illuminate contributions to smooth running in our analysis of the GE gadget, even in cases where those contributions have no design history, so we might also illuminate contributions to fitness that have no history of modification under selection.

Weak reverse-engineering in its standard format in evolutionary biology asks "Why would this part be here?" where a satisfactory answer should provide an account of how the part contributes to fitness. We attribute to an imagined designer of the organism the goal of maximizing fitness, and ask how the parts might contribute to that goal. However, fitness need not be the capacity we fix on. Suppose we are trying to understand a clearly maladaptive phenomenon, such as the spread of cancer cells. Here we might take spreading of cells as the capacity to be analyzed, and we can ask for any part of the system in which cancer cells spread how it might contribute to that capacity. If we proceed under the assumption that all parts contribute to the spreading of cancer cells, we will also increase our chances of teasing out the basis of this capacity.

Weak reverse-engineering is not confined to biological systems in principle. In evolutionary biology we ask how each part might contribute to fitness; in the case of a tumor we ask how each part might contribute to the spread of cancer cells; in the case of a glacier we ask how each part contributes to erosion. The appeal to the thought of a designer behind these capacities adds nothing beyond a spur to the investigator to not assume too quickly that any part of the whole is irrelevant to the capacity in question. However, as a matter of psychological fact the invocation of design talk is far more likely in connection with an eye than with a glacier. That is because biological systems are produced through selection in such a way that they can acquire complex adaptations.

The very generality of weak reverse-engineering should make us wary of thinking that all valuable artifact thinking signals a feather in the cap of adaptationism. Ernst Mayr may be guilty of this when he tells us: "The adaptationist question, 'What is the function of a given structure or organ?' has been for centuries the basis for every advance in physiology" (1983, p. 328). We can make great advances in asking what the function of a structure or organ is, where what we gain is an understanding of the operation of a wholly maladaptive system.

3.3 Adaptationism and the Artifact Model

Weak reverse-engineering, although important, does not account for the ambition that adaptationists have for the artifact model. The value of an engineering approach to organisms is not exhausted by using an imagined designer as a prompt for illuminating subtle or surprising contributions to properties of complex systems.

Adaptationists argue that by thinking of the processes by which lineages change as a design process, we can uncover not only current contributions to complex capacities, but historical facts about organismic lineages and their environments. Specifically, we can uncover which problems organisms were designed to solve, and which solutions are likely to be present

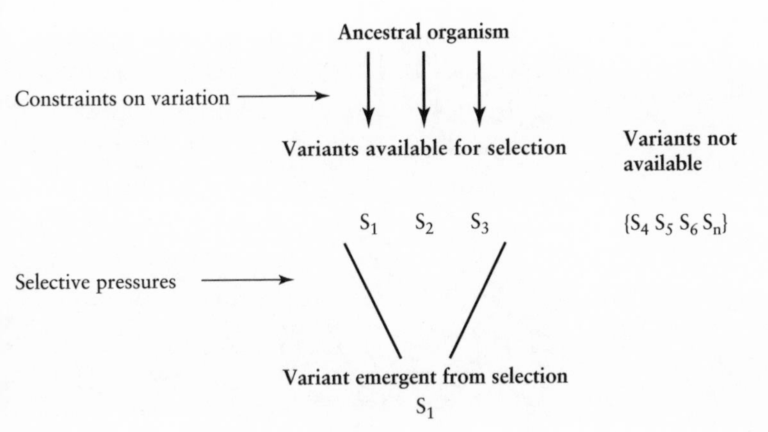

Figure 3.1
The artifact model for organisms.

given those design problems. The next sections present an exposition and critique of this full-blooded artifact thinking.

I begin in this section by sketching models of the processes underlying the production of organisms and artifacts that highlight the similarities between them. The models explain why so many biologists have approached the organic world from the perspective of the artifact model. At the same time, the models themselves show how artifact thinking needs to be handled with care.

The process by which organisms are selected can be represented by the model drawn in figure 3.1, and in figure 3.2 we see a similar representation of the production of a single artifact by an artificer.

The models are isomorphic. In the organic case we can think of a trait that goes to fixation by selection as that member of the set of available candidates that best balances the competing selection pressures that are brought to bear by the environment. Selection pressures are understood in just the same way as they are in chapter 2. In the artifact case we can think of the artifact which is produced as that member of a set of candidate solutions that best balances the competing criteria for choice that are brought to bear by the artificer.

In both cases the set of available solutions is constrained; developmental factors dictate that no gun-toting zebra variant will arise to answer the selection pressure of evading predators, and equally, facts about an

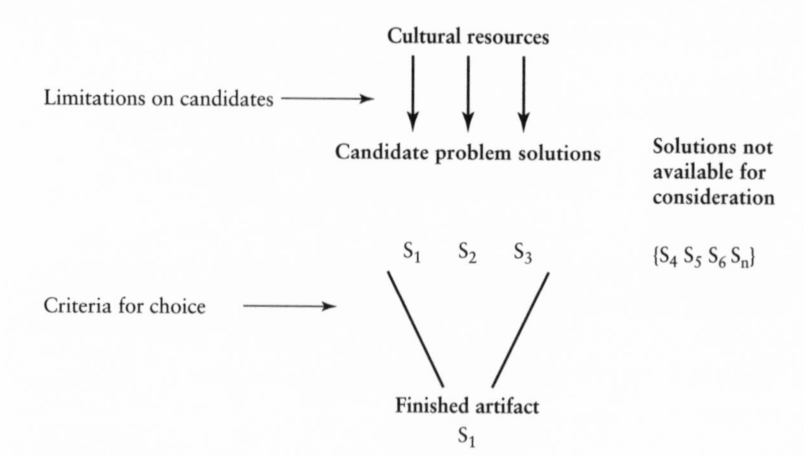

Figure 3.2
The artifact model for artifacts.

individual's cultural environment will dictate that only a handful of possible solutions to a design problem will be considered. Although the extent of constraint may be greater in the organic case, it is surely true that an artificer's cultural heritage and the material resources available will predispose her to ignore some solutions and focus on others. The Greeks would not have thought to use PVC to make furniture coverings.

The models in figures 3.1 and 3.2 make clear that when unraveling the historical processes leading to the emergence of organisms or artifacts we can ask similar questions—for example, "Why is the observed organism/artifact S_1, and not the apparently fitter/better S_2?"—and we can give similar responses. A selection of schematic answers might include:

(I) S_2 was not prevented by any constraint, but never emerged as a variant/was never considered through happenstance.

(II) S_2 would have been fitter/a better solution than S_1 but was prevented from emerging as a variant/being considered through some constraint.

(III) S_2 was in fact less fit/a worse solution than S_1, owing to some unperceived selection pressure/some unperceived nuance to the problem.

(IV) S_2 was present with S_1, and S_2 was fitter/a better solution than S_1, yet S_2 was eliminated from the population through drift/from the artificer's mind by some random event.

Consider, by way of an example of this form of questioning, George Williams's musing on the design of the eye: "Why do we blink with both eyes simultaneously? Why not alternate and replace 95% visual attentiveness with 100%? I can imagine an answer in some sort of trade-off balance. A blink mechanism for both eyes at once may be much simpler and cheaper than one that regularly alternates" (1992, pp. 152–153). Williams's question is exactly the kind suggested by the artifact model; one asks why not some apparently fitter solution S_2, instead of the observed solution S_1? Williams gives a tentative answer that suggests a combination of constraint and trade-off; for all accessible phenotypes with alternating blinking mechanisms, those mechanisms would result in a net loss of fitness because of the trade-off between lower efficiency and increased attention.

Precisely because organisms and artifacts are subject to similar problems in the explanation of their form, an appreciation of the artifact model allows us to highlight the dangers of shallow artifact thinking in biology. Some of the assumptions we may use in the interpretation of form fall short of perfect reliability in both domains. So, for example, when interpreting the design of an artificer, we might assume that he or she always makes the best choices available to them. Hence, when we ask why a painting features one pigment rather than another that is brighter, we might assume either that the brighter pigment was not available to the artist, or that she had some good reason to use the duller pigment. This assumption is not always justified. Sometimes an excellent solution that is available will never be considered, simply through oversight. This is analogous to a phenotype that would be fitter than that which emerges, and which is not prevented from emerging by any constraint, but which nevertheless fails to arise in the population owing to mutational or developmental happenstance.

In sections 3.4 and 3.5 we will see that although the artifact model has limitations, these limitations do not force us to abandon the model—on the contrary they show simply that we have mistakenly applied techniques to the explanation of organisms that we should never have applied even to artifacts. However, before moving on to point out the ways in which thinking about the form of organisms and artifacts overlap, it is important to point out the genuine failings of the artifact model.

First, we can see that the way in which the artifact model is set up discourages recognition of drift. As we saw in chapter 2, drift is not the kind of force that can act on individual entities. In thinking of a trait as analogous to an artifact undergoing shaping over time, we conceive of it as an individual entity subject to the selective forces that work on individual entities. Drift and selection simply do not act on individual entities. For this reason, it is hard to find a strict analogue to drift when we think of the design of an individual artifact.

A very partial analogue to drift is present in the construction of artifacts, as when a painter uses a yellow pigment over a brown one for no reason at all—not because it is cheaper, nor because it facilitates some desired effect more clearly—she simply picks whatever pigment is nearest at hand. This qualifies as a drift explanation in the sense that the choice of one pigment over another is unprincipled, or "random"; the choice of pigment does not reflect the demands of the situation in the same way as those exemplary cases of drift where all the brown horses are killed by lightning strikes, leaving only the black horses to survive; the success of black over brown does not reflect a response to the demands of the environment.

We have seen, however, that drift explanations in evolutionary theory do not require that individual organisms survive or die through some special, stochastic type of cause; rather, the forces that affect individual organisms so as to lead to drift can be identical with the forces that lead to selection. An example will make this clearer.

Suppose an artist always works with pigments at the same distance from her. In the long term, if we assume the artist has a disposition to pick the nearer pigment, the yellow pigment has a higher propensity to be picked than the brown. In this sense we might say that it is fitter. Hence, from the population point of view of evolutionary theory we say that the yellow pigment is selected in virtue of its proximity; this is compatible with there being no reason behind the artist's choice. Each individual selection of yellow over brown occurs "for no reason"; however, from the population perspective yellow is selected—it is at a selective advantage in virtue of its position.

Conversely, if the orientation of the palette with respect to the artist is frequently altered, so that brown and yellow have equal chances of being nearer to the artist in the long run, we can pick out particular sequences in which yellow is used far more often than brown and claim those as sequences showing drift. Yellow has the same fitness as brown here, even though the artist may pick yellow on some occasion quite knowingly. She sees that it is yellow that is nearest, and uses yellow. In this case she selects yellow intentionally, even though yellow is at no long-run selective advantage. There is no paradox here: while natural selection applies to populations, intentional selection applies to individual objects. From the perspective of the individual agent there is a reason for the choice made. From the population perspective, series of choices will show drift.

We can see, then, how conceptualizing evolution in the manner of the criteria of intentional design will tend to lead us to ignore drift, for it leads us to look in the wrong place for it. Even though, in our last example, the artist has reasons for all of her choices, the sequence of choices shows considerable drift. No choice is made at random, all are principled: yet drift is there all the same. This may explain why some population geneticists are hostile to engineering-style approaches to evolutionary explanation. Engineering explanations tend to think of the evolving species in terms of a single archetypal organism on which diverse forces have a shaping role over time; such explanations detract from the statistical nature of population genetic explanations, which explain the relative proportions of traits in populations using the statistical concepts of drift and selection.

A second shortcoming of the artifact model is that it implies that traits are strictly analogous to the parts of artifacts. This, however, exaggerates the independence of organic traits. In chapter 7 I will defend the claim that artifacts themselves evolve by a selection process. Even so, the parts of artifacts often have quite distinctive selective histories. One designer can produce the wheels of a car, another the carburetor, another the chassis. While the parts of organisms have some functional independence in their developmental pathways, they do not evolve independently of each other. Hence the artifact model also encourages us to ignore developmental relations between traits.

Finally, we saw in chapter 2 that the artifact model will miss some of the interests of population geneticists altogether. This is because in some cases the determinants of fitness cannot be identified with selective forces acting on organisms at all. To recap, work by Charlesworth and Giesel (1972), and others, shows the fitness of traits can depend on facts about the population taken as a whole, which cannot be translated into facts about what forces act on individuals within a population.

3.4 Reverse-engineering and Optimality Modeling

Reverse-engineering, in its strong form, seeks to infer adaptive problems—that is, selection pressures—from observed organismic solutions. Paul Griffiths (1996) has criticized this form of thinking on the grounds that observed form underdetermines the nature of the adaptive problems solved. He argues that to be able to make confident predictions of the adaptive problems that a given solution is the response to, we would need to make use of what he calls "functional generalizations." These would be cross-species generalizations which state that any organism, faced with some adaptive problem P, will adopt solution S.

Griffiths is skeptical of the existence of such generalizations, and his reasons drop out from the model outlined in figure 3.1. First, the fitness of a given variant will be dictated not simply by gross features of the external environment, but also by fine-grained features of the environment, the idiosyncrasies of systemic features of the organism, and by population structure. Second, systemic idiosyncrasies also affect which variants are available for selection, hence which solutions are available in a lineage to answer local selective pressures. Third, even when solutions are available to a lineage, when populations are small, or when the selective advantage of the solution is very small, the solution may be lost to drift. So we should be surprised to discover any broad functional generalizations that range across species. Of course such generalizations may exist, but to show that they do so will be an onerous empirical task achievable only on a case by case basis.

Frequently, adaptationists point to cases of adaptive convergence as evidence of reliable functional generalizations that do range across species. Birds and insects evolved wings independently, so it would seem that we can indeed say that, in at least some situations, when the same adaptive problems are faced the same solutions are adopted. The adaptationist is right to say that there are many cases where closely resembling traits emerge independently, in the sense that the resembling traits T* and T** may not owe their resemblance wholly to common descent from some third trait T. In other words, not all resembling traits are homologies. However, this fact alone certainly does not entitle us to infer the existence of broad generalizations that range across all species. Even in instances where common selection pressures explain adaptive convergence, this may be the case only against a background of shared homology. Homologies shared between lineages may affect the adaptive trajectory of those lineages in at least two ways; first, the shared trait may affect the production of variants—that is, it may act as a constraint on variation, thereby affecting the range of candidates available for the solution of adaptive problems. Second, the continued presence of the homology may crucially tip the overall balance of fitness of the available variants, so that a given mutation is likely to yield a variant that will proceed to fixation only when the homology is present.

Suppose then, as seems likely, that broad functional generalizations are rare, and that the most we can hope for are pseudogeneralizations of the form "Faced with adaptive problem P, against a morphological background M, and in populations of type G, adaptive solution S will tend to evolve." Where does this leave adaptationism and the artifact model?

It should be clear that the artifact model remains intact. Nothing in our representations of the forces underlying the production of organisms or artifacts excludes the possibility that the solution adopted in a given environment will be dictated in part by the broad selective pressures that are brought to bear, and in part by fine-grained cultural or phylogenetic norms—either in the form of idiosyncratic local selection pressures, or as constraints on candidate solutions to be selected. However, it clearly is a mistake to assume that we can always make a reliable inference to the best explanation when we attempt to infer past selection forces from the observation of organismic form alone; this is where artifact thinking needs to proceed with caution.

Let us consider the artifact case first. Suppose we are presented with some unusual artifact. If we rely only on data relating to its structure, we would be unsure of the reliability of any inference we might make as to the criteria that dictated its choice and the other candidate solutions with which it competed. For many artifacts there are simply too many combinations of constraints on variation and selective criteria that give both consistent and plausible explanations of observed structure. If we know only that a structure is rigid, flat, and supported by four poles each ten feet high, then we cannot tell if it was designed to be a platform for speech making, a shelter from rain, a shelter from sun, an oilrig's helipad, or a collector of solar energy. If the only data we have to go on are data about the artifact's form, then our hypotheses about its design history will often be underdetermined by those data.

Do we not sometimes read the design history off an artifact quite reliably simply by examining its structure and nothing more? This may be true, but the explicit procedures we go through to determine functions do not reveal the totality of information we bring to bear when we infer the function of an artifact. It is precisely because we often examine items that are products of a design tradition we participate in, that we are able to bring a good deal of implicit knowledge of design constraints and selection pressures to bear when we examine an artifact for the first time. Moreover, artifacts may often carry far richer information about their use than do fossils from their context of discovery—we can construct a picture of the function of an artifact from its surrounding environment, other artifacts discovered with it, and so on. If we know that our flat rigid structure was found on an oilrig, and if we see the large "H" written across it, then we can use this contextual information and our privileged cultural insight regarding the likely signification of the "H" to infer quite safely that the structure is a helipad after all.

In the absence of this kind of additional information, whether it is adduced overtly or covertly through our knowledge of local design traditions, there are simply too many plausible combinations of constraints and selection criteria to make any reliable inference about the problems that an artifact is designed to solve. If we do make such inferences with an exotic artifact, and we make them confidently by using principles only valid for our local design tradition, then we are liable to be led astray. This is precisely the risk run by an evolutionary reverse-engineer who formulates a hypothesis that explains the design of some observed trait, but who does so without information regarding (for example) local constraints on variation, or comparative information regarding habitat.

In case the reader thinks I am inventing these problems in the artifact realm, the caution addressed to optimality modelers in evolutionary biology is also addressed to those investigating artifacts such as paintings. Witness, for example, the discussion of Piero della Francesca's *Baptism of Christ* by the art historian Michael Baxandall. Baxandall's caution has general application in both the realm of organisms and the realm of artifacts:

> Many pictures, including the *Baptism of Christ,* are enclosed in a terrible varnish or carapace of false familiarity which, when we think about them, is difficult to break through. This may be partly a matter of the museum-without-walls syndrome but it is even more a matter of the medium of the art, the fact that most of us are not, at least at this level of accomplishment, skilled executants in the medium . . . A first task in the historical perception of a picture is therefore often that of working through to a realization of quite how alien it and the mind that made it are; only when one has done this is it really possible to move to a genuine sense of its human affinity with us . . . A failure to do this is a main cause of plain historical error. (1985, p. 115)

Baxandall's approach to uncovering the historical processes that result in the production of pictures is a pure project of reverse-engineering. He explains his approach early in his book *Patterns of Intention* using the example of the Forth Bridge; the same project is later applied to paintings: "The sequence began by positing that the object of interest, the Bridge, was a concrete solution to a problem. The solution was in a sense given and visible: the problem was not, except in the guise of a mile of water" (ibid., p. 35). Baxandall thus sets out to uncover, to the best of his abilities, the fine-grained nature of the problems addressed by Benjamin Baker, the designer of the Forth Bridge, and the problems addressed by Piero della Francesca, the designer of the *Baptism of Christ*. To do this, he does not simply bring to bear the data offered by the artifacts themselves. Instead, he adds further information regarding, for example, the range of cultural resources that were available to the designer, and detailed information regarding cultural and other circumstances that might have had an effect on the criteria for choice of one solution over another. If one knows not only the nature of the solution—that is, the form of the artifact—but also which other candidate solutions it might have been in competition with, then one can exclude many hypotheses about the problem the item addresses. Equally, if one can gain contextual information regarding the nature of the culture into which the artifact was introduced, one also sharpens the picture of the likely nature of the problem the item addresses. Without these additional data, hypotheses regarding combinations of problems and constraints are underdetermined.

It is no coincidence that Baxandall's cautions to the investigator of artifacts have strong resonances with Griffiths's (1996) cautions to the adaptationist. Griffiths complains that the adaptationist tends to infer pairs of functional generalizations and historical assumptions from knowledge of the observed trait alone. As we have seen, this move would certainly be suspicious in the artifact realm, where knowledge of an artifact alone typically does not suffice to narrow the range of compatible pairs of problems and constraints. Griffiths thus advocates taking what he calls "the historical turn in the study of adaptation"; that is, he argues that we should attend directly to the historical assumptions made by adaptationist hypotheses in order to narrow the range of hypotheses further.

The upshot of our discussion of reverse-engineering is a warning to adaptationists: they should not suppose that the availability of an adaptive scenario compatible with the observed trait is strongly indicative of the truth of that scenario. However, we should not abandon the artifact model or the spirit of reverse-engineering altogether. Even when reverse-engineering artifacts, we must take a historical turn.

Consider how the need for testing plays out in the most famous example of adaptive explanation of all. Almost all who have noted the increase in frequency in polluted regions of the dark (melanic) forms of peppered moths (*Biston betularia*) since the late 1840s have assumed, naturally, that the melanic form owed its success to camouflaging effects. The trees got darker, and the darker moths prospered—this much is uncontroversial. So surely *crypsis* explains the rise of the darker form. Moreover, Kettlewell's (1973) experiments seem to show conclusively that birds do, indeed, prey on moths according to how well they match the background color of tree trunks. The kind of engineering-type thinking that prompts us to suggest a likely function for wing coloration based on a plausible effect of the wings in an actual environment points strongly in favor of camouflage as the function of the wings. However, subsequent investigators have raised doubts about whether the camouflage story gives a full explanation for the relative frequencies of the various forms of wing. Creed et al. (1980) note that in the industrial Northwest of England, the frequency of the melanic *carbonaria* form remains below 100 percent. Some have posited frequency-dependent selection or heterozygote advantage to explain these discrepancies (see Brakefield 1987 for a short summary). Creed et al. suggest instead that the failure of the melanic form to reach fixation can be accounted for by noting significant differences between the various forms' abilities to reach adulthood—hence nonvisual selection may be a significant factor in explaining observed distributions. Howlett and Majerus (1987) point out that many experimenters derived fitness estimates for the darker and lighter forms by gluing them to tree trunks and recording predation rates; however, peppered moths do not typically reside on tree trunks but instead on the underside of horizontal branches.

Majerus (1998), in a recent book-length study of melanism, concludes that more realistic suppositions of moths' resting behavior in fact favor Kettlewell's own contention that we need not acknowledge significant preadult viability selection to explain industrial melanism. However, Majerus does point out that many models have relied on the reports of human observers to estimate how well-camouflaged the various wing forms are. We now know that birds' visual systems are quite different to humans'. Hence, the observation that birds tend to prey upon moths that are more visually conspicuous to humans does not strongly confirm the proposition that birds tend to prey upon moths that are more visually conspicuous to birds. While Majerus is in no doubt that the primary function of melanism is the avoidance of predation by birds, the more specific claim that the function of melanism is camouflage remains in need of further empirical confirmation. In summary, the confirmation of an apparently indubitable story about the success of the melanic form in industrial areas has required information about, among other things: the visual systems of birds, the behavior of moths, the genetics of the various wing forms, and the historical patterns of variation in frequency of the different forms.

The comments raised about reverse-engineering suggest also that we should interpret optimality models with caution. The structure of an optimality model is easily explained by reference to figure 3.1 above (my account of optimality modeling here is adapted from Sterelny and Griffiths 1999 and Maynard Smith 1978). The inquirer picks some trait or behavior to be investigated—say mammalian gait (Maynard Smith and Savage 1956), the size of a bird's clutch of eggs (Lack 1968) or a creature's foraging behavior (Stephens and Krebs 1986)—and seeks to show how this behavior is optimal relative to some set of constraints and selection pressures. Formally, these models consist of four elements: a fitness measure, a heritability assumption, a phenotype set, and a set of state equations. The phenotype set (or strategy set) describes the variants that are available in an ancestral population for selection to act upon—hence the set is circumscribed by supposed developmental and other constraints. The fitness measure or optimization criterion specifies the currency by which the members of the phenotype set are evaluated. Although ideally one would assign reproductive fitnesses directly to each variant, the model might instead posit some proxy for fitness—assuming, for example in the case of leg length, that the fittest phenotype will be that one which results in the lowest amount of energy needed to cover some distance at a fixed velocity. Clearly one's choice of proxy for fitness must be dictated by one's general perception of the selective pressures at play on ancestral variants, and by one's perception of how traits might interact with each other to affect overall fitness. The heritability assumption reflects the extent to which offspring will resemble their parents in respect of the trait under examination. Lastly, the state equations are a set of rules, often drawn from biophysics or physiology, which order members of the phenotype set in terms of the fitness proxy specified by the fitness measure.

To say that a phenotype is optimal in these models is to say that it is the fittest of those variants that were available for selection. When an optimality model is constructed for a trait, the modeler posits a set of variants (expressed in terms of the members of the phenotype set) and selective pressures (expressed in terms of the fitness measure) that will show whether the observed trait is the fittest of the members of the hypothesized phenotype set. In fact, optimality models play two rather different roles—roles that are expressed through how the model is interpreted when it suggests that the observed trait is suboptimal. For some modelers, this is evidence that selection has not acted to shape the trait in question. For others, this shows only that some element of the model may be mistaken. In the first role, optimality models are used to demonstrate that a trait is optimal, hence that it is an adaptation. If the trait is locally optimal, then this is taken as good evidence for the action of selection on the trait; in other words, if the trait is well designed, then, it is claimed, this should increase our confidence that it is a product of selection.

On a second interpretation, optimality models are used not to demonstrate *that* a trait is an adaptation, but rather to demonstrate *how* a trait is an adaptation. Here it is assumed that the trait is optimal (that is, the fittest of ancestral phenotypes) and the model is used to determine what constraints must have been present, and what selection pressures must have been at work for this to be the case (Parker and Maynard Smith 1990). When used for this function, the name "optimality model" is liable to mislead. As Parker and Maynard Smith say (ibid.), the proposition that natural selection optimizes is taken as a given; it is not what is tested by the model.

Stephens and Krebs are clear about this role for optimality modeling (1986, p. 212): "Even if they serve no other purpose, well-formulated design models are needed to identify constraints: without a design hypothesis there would be no basis for postulating any kind of constraint!" Parker and Maynard Smith are perfectly clear about what we should do with a model if it tells us that an observed trait is suboptimal: "If they fit, then the model may really reflect the forces that have moulded the adaptation. If they do not, we may have misidentified the strategy set, or the optimization criterion, or the payoffs; or the phenomena we have chosen may no longer be adaptive. By reworking our assumptions, we modify our model and revise and retest the predictions" (Parker and Maynard Smith 1990, p. 29). The goal of the model is not to prove anything about the power of selection—instead it is to locate the specific constraints and adaptive forces that have resulted in the selection of the trait in question.

Our earlier discussion of reverse-engineering alerted us to the possibility that a given trait may be compatible with many different combinations of selection pressures and constraints, so that same caution should be exercised with respect to optimality. There may be many combinations of phenotype sets, heritability assumptions, fitness measures and state equations compatible with the claim that a phenotype is the fittest of available variants. Moreover, there may be combinations of these factors that would show that any imaginable trait is an optimal adaptation—even traits that seem adaptively neutral or maladaptive. To take an extreme example, Darwin once claimed (1996, p. 163) that if he ever came across a creature which bore traits that clearly aided some other species—for example, a horse with an organic saddle—then he would give up his theory of natural selection and revert to some form of creationism. But Dawkins's (1982) theory of the extended phenotype now suggests that adaptations that benefit an organism other than that which bears the adaptation can arise through natural selection, without any intervention from a conscious designer. The additional resources of evolutionary game theory, sexual selection, and the general acknowledgement that selection pressures can be exceptionally subtle and unintuitive to human investigators mean that adaptive scenarios, hence optimality models, can be constructed for almost any trait.

Adaptationists are fond of citing "Orgel's Second Law: Evolution Is Cleverer Than You" (Dennett 1995, p. 74). The price that one must pay for pointing out the great cunning of Mother Nature, is to acknowledge also that optimality models which aim to reconstruct her extreme cunning solely from data relating to products of that cunning face a genuine problem of underdetermination. If adaptive histories can be readily constructed for so many traits, then it seems likely that not one but several histories can be constructed for the same trait. If this is indeed the case, then optimality models should be regarded as useful, but only useful insofar as they generate possible histories for a trait. One would want more data before accepting the truth of any model consistent with a trait. As we saw when discussing reverse-engineering in general, one would ideally try to confirm as far as possible the hypothesized constraints, selection pressures, and other assumptions contained within the model. So optimality models are useful, but should be treated with caution. This conclusion is almost identical with that of Oster and Wilson (1984, p. 273): "... optimization models are a method for organizing empirical evidence, making educated guesses as to how evolution might have proceeded, and suggesting avenues for further empirical research."

3.5 Adaptive Thinking

Similar cautions apply to a second form of adaptationist thinking that derives from the artifact model—what Griffiths calls "adaptive thinking." Reverse-engineering seeks to infer the adaptive problem faced from the nature of the solution adopted; adaptive thinking is reasoning that moves in the opposite direction. One tries to infer the nature of the solution adopted from knowledge of the problem faced.

We should begin by noting that there is a counterpart "weak adaptive thinking" to the weak reverse-engineering described in section 3.3. If one understands by "problem" no more than some function that an organism does in fact carry out, then one can ask what the solution might be in the quite simple sense that one can ask, given that the organism achieves P, how it does so. This very general form of thinking can be applied in all sciences: we can regard a complex capacity of a system as a "problem" and try to discover how the system "solves" that problem. What is more, we can think how we might solve that problem ourselves, and use this as a hypothesis for how the same effect is achieved in the system. Such thinking may have heuristic value, but it is no more than a prompt by which one might generate possible processes underlying any complex capacity that one seeks to understand. One could ask, for example, how one would go about designing a machine made largely of ice, and which is able to produce a certain kind of erosion pattern on alpine mountains.

This kind of thinking may well give us an insight into how glacial erosion occurs; however, this kind of problem-based thinking lacks the dynamic, creative elements in nature's problem solving—we assume that over time, so long as the adaptive problems posed remain the same, they can cause their solutions to come into existence in the senses explained in chapter 2. That said, weak adaptive thinking must be mentioned, because many examples of adaptive thinking that are cited in support of the adaptationist program are, in fact, consistent with the kind of benefits that one would expect from weak adaptive thinking alone. Consider the following example from Rüdinger Wehner:

> While foraging in a circuitous way over distances of more than 200 m, *Cataglyphis* ants of the Sahara desert navigate by path integration. They continually measure all angles steered and all distances covered, and integrate these angular and linear components of movement into a continually updated vector always pointing home. This is a computationally demanding task which *Cataglyphis* must solve with its small nervous system, and—as recent research has shown [e.g., Wehner et al. 1996]—it does this by relying on a number of rather simple subroutines. (Wehner 1997, p. 33)

If Wehner means to assert here only that *Cataglyphis* is able to navigate across large distances, and that this is achieved using a small nervous system, hence that one must explain this capacity without positing very powerful computational resources, then the talk of "tasks solved" is compatible with weak adaptive thinking.

Still, the appeal to a stronger form of adaptive thinking, and one tha typically makes explicit reference to artifacts, is often made by evolu tionary biologists, and by evolutionary psychologists in particular. Kreb and Davies make a general case for this approach, and the evolutionar psychologists' case is presented by Barkow, Cosmides, and Tooby:

> Visitors from another planet would find it easier to discover how an artificia object, such as a car, works if they first knew what it was for. In the same way physiologists are better able to analyse the mechanisms underlying behaviour onc they appreciate the selective pressures which have influenced its function. (Kreb and Davies 1997, p. 15)

> By understanding the selection processes that our hominid ancestors faced—b understanding what kind of adaptive problems they had to solve—one shoul be able to gain some insight into the design of the information-processing mech anisms that evolved to solve these problems. (Barkow, Cosmides, and Toob 1992, p. 9)

The analogy with artifacts is appropriate, partly because it again ex poses the limitations of crude adaptive thinking. If we tell Krebs an Davies's alien visitors only that the object they are confronted with i an object designed for transporting people and their shopping, then the will have open to them a vast number of hypotheses for how that prob lem might be solved—perhaps a car, but perhaps roller skates, a canoe, a horse, or a spaceship. What is more, they will miss out on the many func tions of a car that are not contained within the phrase "item designe for transporting people and their shopping." Unless they are told of som of the car's additional functions, they will not predict that it is painte bright red and shaped in an anatomically suggestive way, nor that it bear numerous small "Ferrari" badges.

If all we know of an artifact is the broad problems that it was designed t solve, then this will tell us very little about its structure or inner workings. As the model sketched in figure 3.2 makes clear, adaptive thinking abou artifacts will only bear fruit if we can add to our knowledge of broa problems faced further knowledge of the design tradition that generate candidate solutions and knowledge of the more fine-grained problems an cultural preferences that affect the final selection.

Of course in the example of the car, our alien visitors are not (w imagine) presented with a screen and told that behind it there is som item designed for transportation. They can also see and examine the ca itself, so that the project of uncovering how it works involves an interplay between observations of what materials it is made from, various effects when the steering wheel is turned or the clutch is depressed, together with hypotheses about what parts one might expect to find in an object of that sort, on the hypothesis that it has been designed for this function. Many of their hypotheses (that the transportation device is a pair of roller skates, for example) can be immediately ruled out in the face of these observations.

Evolutionary psychologists favor adaptive thinking in part because they feel that inferring the mind's structure and mechanism from the problems it was designed to solve can elude the methodological difficulties presented by the fact that the mind, unlike the car, cannot be so easily tinkered with and examined directly. Still, most evolutionary psychologists agree that if an evolutionarily anticipated mechanism does not appear to be present by the lights of the conventional methods of cognitive psychology, then the suggested adaptive problem either never existed, or was never solved: "... nature always gets the last word. For example, it would appear to make excellent adaptive sense for human males to be able to detect female ovulation and to find ovulating females most sexually attractive; but the preponderance of evidence is that no such adaptation exists" (Symons 1992, p. 144).

There is no question that adaptive thinking has some value; however, just as reverse-engineering is a risky business when the only datum we have is the trait in question, so adaptive thinking is a risky business when all the data we have are broad problems suggested by scant knowledge of an ancestral environment. Although adaptive thinking can provide us with an engine for the generation of hypotheses to test, we should be extremely wary of according any significant credence to those hypotheses in the absence of such testing and observation.

There are good reasons to be especially wary of the use of adaptive thinking in evolutionary psychology (see Lewens forthcoming-a for expansion on these claims). Unless we have quite detailed information regarding likely developmental constraints on cognitive variation in ancestral environments, as well as information relating available variation to overall fitness, then we should expect no very accurate predictions from thinking based on environmental problems. Adaptive thinking in evolutionary psychology is likely to be successful only if we already have a rich store of information regarding the development and organization of the mind; however, it is precisely this information that adaptive thinking is supposed to deliver.

Cosmides and Tooby point out that we do have access to a rich body of information about our ancestral environment: "Our ancestors nursed, had two sexes, hunted, gathered, chose mates, used tools, had colour vision, bled when wounded, were predated upon, were subject to viral infections, were incapacitated from injuries, had deleterious recessives and so were subject to inbreeding depression if they mated with siblings, fought with each other, lived in a biotic environment with felids, snakes and plant toxins, etc." (Cosmides and Tooby 1997).

All these things are true, but still a central problem remains in deciding how to translate a description of a physical environment into a set of predictive adaptive problems. To take the case of plant toxins, should our ancestors have evolved a general horror for novel plant species? Should they, instead, have evolved means of removing the toxins from plants by cooking them? Should they have evolved specific and quite different horrors of varying intensities for different types of toxic plants? Instead might we expect a wholly general cognitive module that gives us fear of the unknown, thereby ensuring that we do not go near potentially dangerous creatures, plants or strangers? All of these seem to be suggested by the environment in which our ancestors evolved. Now perhaps the evolutionary psychologist will think that all of these are worthy of testing. And perhaps some tests will show that the adaptation in question does exist. In this case, we can grant some predictive success to adaptive thinking, but note that it is against a general background of failure.

Again, all of these cautions and limitations on adaptive thinking are predicted by the artifact model. The question of how much help we can expect from broad problem-based thinking for uncovering how a car works might be met by a number of responses from our alien engineers. Some will argue that there must be some dedicated mechanism within the car designed to solve the problem of parallel parking; others will say that although there is evidence that cars are able to be involved in parallel parking, there is no firm evidence to suppose that anything more than a general purpose "central driving" module facilitates this.

In a similar debate within evolutionary psychology, Cosmides and Tooby (e.g., Cosmides 1989) argue that the needs of our ancestral environment predict that we should have evolved a dedicated "cheat detection module." Others (see, for example, Over forthcoming) argue that although adaptive thinking might suggest that we ought to have evolved a cheat-detection model, the weight of empirical evidence from direct testing of our mental abilities is not sufficient to overturn the hypothesis that detecting cheats is achieved using a general purpose reasoning module. What the artifact model recommends for the investigation of both organisms and artifacts is the accumulation of as much data as possible on the structure and behavior of the item to be investigated, the fine-grained selection pressures (criteria for choice) in the environment in which it was formed, and an understanding of the range of variants with which it had to compete. Reverse-engineering and adaptive thinking are both weakened the more elements of this equation are removed.

3.6 Adaptationism Revisited

My strategy in sections 3.4 and 3.5 was to show how many problems raised for engineering-type interpretations of organisms are also present when we try to interpret artifacts. Although we have seen that inquiry in both domains faces problems of the same type, we have not shown that these problems exist to the same degree in both domains. We need access to similar bodies of data—data regarding constraints on the availability of solutions and fine-grained selection pressures—when we try to infer the problems solved by an organism or artifact. However, we saw in section 3.4 how our participation in local traditions of artifact design may give us improved access to these types of data when we come to think about artifacts that are also part of these traditions. This does not undermine the artifact model itself; it shows only that artifact thinking about organisms may face greater practical problems when it comes to the acquisition of data than artifact thinking about artifacts.

What does this tell us about the debate over adaptationism? A conception of adaptationism as nothing more than the view that one should investigate organisms through the lens of the artifact model is such a weak position as to encompass most anti-adaptationist concerns. As we have seen, many traditional criticisms of adaptationism—that it ignores constraints and drift, say—and even more recent criticisms like those of Griffiths about the dangers of simpleminded reverse-engineering or adaptive thinking, can be phrased within the framework of the artifact model. Even if we agree that drift never plays a role, then the artifact model still allows that organisms as a whole may be nothing like good design solutions owing to complex trade-offs, interactions between traits, and constraints. This all helps to explain why the debate over adaptationism is sometimes so hard to keep a grip on, with both sides attacking each other while seeming to agree over so much. To caution against bad artifact thinking is not to reject artifact thinking altogether. Dennett's (1995) adaptationism—thought by many to be extreme—is not so extreme after all.

We saw in section 3.3 that the artifact model can discourage the generation of hypotheses that mention drift. Adaptationists frequently acknowledge its action in their methodological discussions; however, it is rare to find drift seriously entertained as a possible explanation for the emergence of some trait, or for the failure of some imagined competing trait to emerge in a population. That said, what we might call hypothetical anti-adaptationism—the claim that drift has played a significant role in the evolutionary process—is, I suspect, only a stepping stone to a rather different epistemological grumble of many who call themselves anti-adaptationists. Their claim is not that adaptationists fail to recognize drift in specific cases when they should, but instead, that the existence of drift makes it difficult to say just which selective forces have acted on a trait. It is important to note here that similar epistemological worries would apply if drift had no significant role in evolutionary explanation. Even then, it may be that we do not have sufficient data to say just which selective forces and constraints gave rise to the forms we see today or in fossils.

"Anti-adaptationist" is a misleading name for those who hold that our data about evolutionary history are so scant that we should be wary of accepting particular adaptationist hypotheses. "Evolutionary skeptic" would perhaps be a better name, for the reasoning that leads one to reject hypotheses of selective origins as poorly supported by data should also lead one to reject drift hypotheses as poorly supported by data.

"Just-so stories" (Gould and Lewontin 1979) can be told about drift as much as about selection. So for the evolutionary skeptic, whether evolutionary historical hypotheses refer to drift or selection is irrelevant. This suggests in turn that the biologist who pits herself against the evolutionary skeptic should not be thought of as holding that selection is in some sense the most important evolutionary force, but instead to the thought that our methods are good enough to reconstruct evolutionary histories. I suspect that many today who style themselves adaptationists or anti-adaptationists are divided not so much on what we might call the metaphysics of evolution, but its epistemology. Focusing on charges of Panglossianism obscures this important difference and delays its resolution. Many of the central debates over what has been called adaptationism should be resolved by investigating, and if need be by strengthening, our methods for ranking evolutionary hypotheses, and for obtaining data regarding the evolutionary past.

This chapter has left several issues outstanding. First, there is more to be said about the artifact model. Some critics have argued against it in a radical way, claiming that the very concept of an evolutionary problem is misconceived. Others say that we should replace the teleological form of inquiry prompted by the artifact model with a purely "structural" approach. These attacks on the foundations of the artifact model are considered in the next chapter.

4

On Five "-Isms"

4.1 Attacking the Foundations

The last chapter focused primarily on epistemological limitations of the artifact model. There we looked at whether reverse-engineering and adaptive thinking are likely to give us strong guides to the selective problems that faced organisms in the past, or to the solutions that have emerged to cope with those problems. In this chapter I want to look at some more basic critiques of the artifact model that threaten the very concepts of an adaptive problem, of adaptation, and of the teleological approach to understanding natural design. We will need to examine the relationship between the artifact model and five *"-isms": adaptationism, developmentalism, constructivism, internalism,* and *externalism.*

Developmentalism accuses *adaptationism* of the mistake of concentrating exclusively on questions about the function of organic traits, at the expense of inquiry regarding their form. A particularly strong example of this kind of skepticism, to which we will return later, is voiced by those who defend a radical structuralist school (Ho and Saunders 1984): "Our common goal is to explain evolution everywhere by necessity and mechanism with the least possible appeal to the contingent and the teleological. Accidental variation and selective advantage . . . are thus relegated here to the last resort."

Constructivism (Lewontin 1984, 1985), on the other hand, holds that the concept of adaptation is ill formed. Lewontin complains that organisms do not adapt to fit problems laid down by some prespecified niche; rather, the activities of organisms result in the construction of those niches, with the result that the adaptationist's conception of an environmental problem is indefensible.

Although I am sympathetic to many elements of the developmentalist and constructivist critiques, I argue that the terms in which they are phrased suggest a mistaken opposition between developmental and adaptationist thinking, and that a workable concept of an "adaptive problem" can be salvaged in spite of the acknowledged construction of environments by organisms. Making good this claim demands discussion of the two remaining *-isms*—Godfrey-Smith's (1996) *internalism* and *externalism.*

4.2 Selection and Development

In chapter 2 I argued that although selective forces explain adaptation, they do not do so alone. The right kind of developmental organization is needed also, if reproducing systems are to be apt for cumulative evolution. This means that if our goal is the quite general one of explaining how adaptation is possible, then answers will be required both from those whom we might call traditional adaptationists, who typically stress the importance of "external" factors like selection in shaping form, and those whom we might call "developmentalists" who look to internal facts about developmental programs and developmental constraints.

This aspect of the relationship between adaptationism and developmentalism suggests that the study of development will inform the concerns of adaptationists only when we consider the very broad question of the generic developmental organization required for adaptation. Outside of this arena, we might expect the two groups to work in isolation from each other, pursuing their different explanatory interests. On such a model, there should be no true conflict between adaptationists and developmentalists. Rather, once the groups move away from the shared *explanandum* of how adaptation is possible, any apparent conflict reflects nothing more than a difference of explanatory interests (Amundson 2001 makes the same suggestion). Developmentalists, on this understanding, are interested primarily in explaining the nature of developmental programs—what phenotypes they permit, what phenotypes they do not permit—while adaptationists take more of an interest in asking how those phenotypes can be expected to vary in fitness.

In fact, developmental work can inform day-to-day adaptationist concerns directly, by providing important information for the testing of adaptationist hypotheses. Confronted with a population of light green moths residing against a somewhat darker green background, one might ask why we do not observe moths that are a better color match. Now the adaptationist instinct is to ask why a darker colored moth would, in fact, have been less fit than the moths we see.

Here the adaptationist could propose a trade-off. Had there been a darker moth, it would have been less fit owing, perhaps, to diminished visibility by potential mates. The adaptationist tends to tell a certain kind of story—there were such darker green moths, but they were less fit than the light green ones. Yet it may turn out that we do not see dark green moths because the developmental program of the moth does not permit the synthesis of such a pigment. If this is the case, such dark green moths are purely hypothetical—they never entered into competition with the light green ones.

In this scenario, discoveries from developmental biology would be used to confirm adaptive hypotheses, and adaptive studies might suggest likely avenues for research by developmentalists. Direct developmental experimentation would help determine what constraints, hence what likely variants, might have existed in ancestral populations. They would tell us whether the suggested trade-off between camouflage and recognition is needed to explain the nonappearance of a better camouflaged moth. Maybe it will turn out that constraint prevents the darker variant from emerging, and no trade-off is needed to explain why the better camouflaged moth is not present. Adaptive predictions that are not met might also suggest some likely constraints that developmental biologists might examine. The failure of color to match perfectly might suggest the existence of a constraint that prevents the better match emerging.

In practice, there is not such a friendly relationship between adaptationists and developmentalists. There are genuine disagreements between both sides about matters of fact, not just divergences in primary explanatory interests. The two camps tend to disagree, for example, in their assessment of how wide the range of variation is (Sober 1998). The adaptationist tends to assume that the developmental programs of existing creatures allow a wider range of viable phenotypes to emerge than the developmentalist does. This, I think, lies at the root of the idea that adaptationism is in conflict with developmentalism, in that the former stresses the "power of selection," while the latter stresses the inability of selection to overcome constraint. However, this form of adaptationism is not so much a thesis about the ability of one force to overwhelm other forces; rather, the strand of adaptationism that stresses the broad range of available variation is itself a thesis about the plasticity and independence of developmental programs.

Adaptationists and developmentalists do disagree, but just how to express that disagreement is underdetermined. We just saw that one way of seeing the point of disagreement is to locate it in the nature of development, but the same set of facts, and the same disagreement, can frequently be expressed as a debate over the nature of salient selection pressures. The small set of available phenotypes acknowledged by the developmentalist might be explained in terms of early-acting selection processes. In the example of the moths, the developmentalist might underwrite the appeal to a constraint that prevents the emergence of darker-winged forms by showing that developing moths that began to lay foundations for darker wings would encounter some large developmental problem early in ontogeny, and that they would tend to abort as a result. This way of phrasing what is at stake does not leave one camp emphasizing selection whereas the other emphasizes development; instead, one camp assumes that a large number of conceivable adult phenotypes could compete with one another whereas the other camp assumes that most selection occurs early in ontogeny. Again, the debate between the two camps need not be understood as one between those who stress the power of selection and those who stress the power of development.

Biologists like Reeve and Sherman (1993, 2001) who defend adaptationism by including such cases of embryonic selection in their models have already absorbed into their view many of the most important points made by earlier developmentalist critiques of the adaptationist program (e.g., Gould and Lewontin 1979). By acknowledging that many significant selection processes occur in virtue of interactions between the parts of the developing organism, these adaptationists' attention is reoriented away from local environmental features, and toward features of developmental processes that might be common to many species whose external environments are quite different.

Where, finally, does all this leave the artifact model? Developmental critiques point out quite legitimately that if one focuses only on the relative performance of different variants with regard to features of the external environment, then one risks ignoring a range of salient facts that explain the appearance, and nonappearance, of certain forms in terms of the demands of internal developmental processes. So the developmental critique does threaten any naive artifact model that implies we might discover all there is of interest in the explanation of evolutionary change by looking to the problems laid down by external environments. But a more sophisticated artifact model has two ways of acknowledging the importance of attention to development: the first is through constraint, and the second is by seeing developmental processes as solutions to problems laid down by other parts of the developing organism. The second way involves a kind of complexification of artifact thinking to recognize internal parts of an organism as environments for other parts, and allows engineering-style analyses to be applied once more in ways that necessitate detailed attention to developmental processes and their interactions. Artifact thinking that recognizes only features of the external environment as laying down problems will tend to discourage such close attention to development, in favor of black boxing the details of developmental processes and referring instead to constraint. In neither case should attention to development cause us to doubt whether the artifact model is appropriate at all. There is, however, another way in which developmental thinking could render artifact thinking unattractive, but I will wait until chapter 6 before I explain how.

4.3 Gradualism and Goodwin

According to a fairly widely accepted definition, a trait is an adaptation for some function if, and only if, that trait has been selected for that function. Sober's definition is typical: "A is an adaptation for task T in a population P if and only if A became prevalent in P because there was selection for A, where the selective advantage of A was due to the fact that A helped perform task T" (1984a, p. 208).

This is in some ways a curious definition, since it fails to accommodate the intuition that an adaptation is some trait that has been "shaped" by selection for its function. To be selected for a function is simply to make a contribution to fitness that causes organisms with that trait to increase their frequency. On this definition, an eye is an adaptation for seeing regardless of whether the first eye sprang into existence fully formed, or whether there was some progression from partial eyes through many intermediate forms to a fully functional eye. Intuitively, the metaphor of a trait being "shaped by selection for its function" applies in the second case but not the first.

We can already go some way to making this idea of "shaping" a trait more precise. Selection explains the emergence of traits when it increases the chances of otherwise unlikely traits arising. Suppose a population of eyeless organisms has a high chance of developing a complex, functional eye through a single macromutation. Suppose, moreover, that a variant with a partial eye has a very slightly higher chance of developing the same complex, functional eye through a single macromutation. This second partial-eyed variant arises, and selective forces cause it to spread through the population. Here selection increases the chances of a full eye arising, but it increases those chances only a little in comparison to the chance of an eye arising through macromutation from an eyeless variant in the absence of selection. Selection explains how eyes come to be, but it does little explanatory work. It is only when the chances of single macromutations are low that selection itself does enough of this work for the idea that it "shapes" the trait to have force.

It is the idea that most adaptive traits have been shaped by selection, not hypotheses about the selective value of those traits, that attracts criticism from another kind of developmentalist—the process structuralist (see also Smith 1992 for a more detailed discussion of structuralism). Brian Goodwin (1994, p. 148), for example, suggests that the role of selection may be overstated even for complex traits like the eye: "... eyes are not improbable at all. The basic processes of animal morphogenesis lead in a perfectly natural way to the fundamental structure of the eye." A few pages later Goodwin makes the same point in a vocabulary reminiscent of Stuart Kauffman:

> The processes involved are robust, high probability spatial transformations of developing tissues, not highly improbable states that depend upon a precise specification of parameter values (a specific genetic program). The latter is described by a fitness landscape with a narrow peak, corresponding to a functional eye, in a large space of possible nonfunctional (low fitness) forms. Such a system is not

robust: the fitness peak will tend to melt under random genetic mutation, natural selection being too weak a force to stabilize a genetic program that guides morphogenesis to an improbable functional goal. The alternative is to propose that there is a large range of parameter values in morphogenetic space that results in a functional visual system: i.e., eyes have arisen independently many times in evolution because they are natural, robust results of morphogenetic processes. (Ibid., p. 154)

What, exactly, is Goodwin denying here, that the adaptationist would have us believe? He is downplaying the role of selection in favor of developmental processes, but in what sense? Certainly there is no denial that variants with eyes may be fitter than variants without. So in this sense, eyes are adaptations and they are explained by selection. Instead, Goodwin denies that mutations leading directly to eyes are as improbable as one might think. That is, Goodwin appears to deny what adaptationists take for granted—the principle that eyes can only have arisen through a series of small mutations.

In fact, the distinction is not so clear cut. At least some mutations arise spontaneously that increase fitness; the adaptationist merely disagrees about how much of an increase in fitness and what kind of an increase in fitness is likely to come about through this kind of mutation. So the adaptationist condones mutations without intermediaries when they improve the functioning of the eye a very little, but not when they improve function a lot. This move is justified partly on observational grounds—many traits show evidence of change by accumulation of many small mutations rather than a few large mutations—and partly on the theoretical grounds that spontaneous large increments in fitness do not occur, or are exceptionally unlikely to occur. But then Goodwin says at one point that the kind of proto-eye that he has in mind—namely a partially transparent epidermis covering excitable cells—is a result of "extensions and refinements of basic morphogenetic movements" (ibid., p. 154). Here it looks as though Goodwin thinks of these "extensions and refinements" as a series of gradual changes of the sort that would gladden an adaptationist, and few adaptationists will deny that selection works on variations generated around existing developmental programs. Goodwin's complaint against adaptationism is ambiguous. On a conservative reading he is claiming that eyes arise as gradual modifications from "basic morphogenetic movements." In this sense he can be a gradualist about eyes, while attributing the independent evolution of eyes to the nature of these underlying morphogenetic movements. On a more radical reading he is claiming that proto-eyes are likely to arise through just a few mutations, and that we do not need to posit a whole series of small mutations to explain their presence.

Here, again, the debate between adaptationists and developmentalists is typically framed in terms of the question of whether selection, or development, has the primary role in explaining form. On both readings of Goodwin's claim we can make sense of the idea that selection has a diminished role in explaining the emergence of eyes. On the radical reading, the idea would be that some mutations that give a large increase in fitness are really far more likely to arise spontaneously than one might think. If that is the case, then selection retains its role in explaining how many fit traits come to fixation, but loses its role in explaining how those traits are "shaped." So on this radical reading the debate between adaptationists and developmentalists turns into a debate about the likelihood of fitness-enhancing mutations. Again, what looks like a debate about the power of selection turns out to be a debate about the nature of development. The developmentalist might try to show experimentally that at least some fitness-enhancing saltations can arise quite naturally from existing developmental programs. The adaptationist might begin with the premise that a fitness enhancing trait could not have come about by macromutation, hence that a smooth series of adaptive modifications is necessary if the emergence of the trait is to be explained at all. Since both camps are committed to views about the nature of development and adaptive trajectories, I think we should see both as providing important data about the *explananda* of interest to either side. Where they differ is in their starting points. While the adaptationist assumes that fitness enhancing saltations are unlikely, hence that developmental gradualism must be true, the developmentalist takes the probability of fitness enhancing saltations to be an open question, to be answered by empirical investigation.

On the more conservative reading of Goodwin's claim, selection again has less explanatory work to do than traditional Darwinism might suggest, simply because most morphogenetic fields are already such as to facilitate the gradual evolution of eyes. The idea is that in the absence of selection, the random generation of an eye is not likely, but it is considerably more likely than we might have thought otherwise. It is as though the adaptationist asserts that selection builds a 747 from pieces found in a junkyard, while the developmentalist holds that selection builds a 747 from parts found around a 747 production line. The chances of a 747 being formed at random by a storm in a 747 factory, although extremely low, are considerably higher than the chances of a 747 being formed at random by a storm in a junkyard.

The process structuralist tells us, in part, that developmental processes provide all the main building blocks for eyes, and all selection has to do is see to it that the adaptive pathway is followed. In this sense, the claim by the structuralist is a version of the developmentalist claim that variation is highly constrained, with the added claim that it is shared constraints, just as much as shared selection pressures, that explain convergent evolution. That will not be news for many adaptationists. Many will expect developmental programs and the constraints that accompany them to be shared across species owing either to common descent or to convergent selection.

What distinguishes the structuralist approach, and what makes it truly radical, is the claim that only a small number of developmental programs are physically possible, and that it is this fact of physical possibility, not shared ancestry or convergent selection, that explains their appearance across distinct species. For the process structuralist, organisms share morphogenetic fields because these fields are the only ones permitted by univeral "laws of form." All natural examples of diamond have a tetrahedral molecular structure in common, yet we explain this similarity by looking to universal physical or chemical laws that explain the stability of such structures under broadly varying local circumstances. We do not explain similarities in token diamond structures by looking to common descent from a diamond that had this structure, nor to idiosyncratic demands of the local environments of diamond tokens. The process structuralists advocate broadening the ambit of such explanations of common structure to biological traits. I take it that this is what Ho and Saunders (1984) have in mind in the passage I quoted at the beginning of the chapter: "Our common goal is to explain evolution everywhere by necessity and mechanism with the least possible appeal to the contingent and the teleological."

A successful process structuralism could compromise the artifact model by limiting the range of engineering-style explanations for organic forms.

It will not undercut the claim that the forms we see have been selected for, and will most likely not undercut the use of many optimality models in the manner outlined in the last chapter. But the artifact model also encourages us to conceive of the process of selection and mutation as a design process akin to the repeated modification and testing of a range of prototypes by an artisan or inventor. The metaphor of selection as a shaper and refiner of organic form to yield adaptation suggests a model whereby selective forces have primacy in explaining the emergence of adaptation through a process akin to the whittling of an otherwise orderless piece of wood. We may, however, discover that selection does far less "shaping" than our intuitive appreciation of adaptation inclines us to think. What is more, our insistence on engineering explanations that assume the low probability of certain traits arising in the absence of selection may blind us to the true facts about how more general laws govern the probabilities of mutations of various types.

None of the preceding argument establishes the existence of any such laws, or shows any adaptationist hypothesis to be false. My intention is merely to make clear what the aims of the structuralist program are, and how it could come into conflict with traditional adaptationism on the question of whether traits are shaped by selection. It would be unwise for an adaptationist to argue against the effort to carry out the structuralist program. As we saw in the last chapter, many adaptationists are best understood as backing a certain heuristic—a method for investigating nature that will help us to understand how forms vary in fitness, what constraints may have prevented alternatives from emerging, and how selective forces have shaped the forms we see. The question of the role of general laws in explaining the repeated emergence of certain forms cannot be settled a priori; it is an empirical question whether there may be laws that prevent some structures from emerging and encourage others to do so. There is, then, a role for an alternative structuralist heuristic that ignores the characteristic forms of question encouraged by the artifact model, preferring to look directly at the physical principles that underlie development. Such a program may frequently find nothing, but so too will the adaptationist program conclude on many occasions that shaping under selection is not the right way to understand the form of particular traits. Pluralism at the level of heuristics is the most attractive option, and I can do no better than to close this section by quoting D'Arcy Thompson: "Still all the while, like warp and woof, mechanism and teleology are interwoven together, and we must not cleave to the one nor despise the other; for their union is rooted in the very nature of totality" (1961, p. 5).

4.4 Adaptation and Construction

In this section I consider the challenge to the artifact model from Lewontin, who argues that the very concept of an adaptive problem cannot stand up to scrutiny. While I agree with many of his substantive claims about the role of construction in shaping selective environments, I try to show that a concept of an adaptive problem that is suitable to the needs of the artifact model can be salvaged. Before this project is complete it will be useful to learn from Godfrey-Smith's (1996) discussion of two other *-isms:* internalism and externalism.

4.4.1 Internalism and Externalism

Peter Godfrey-Smith (1996) has expressed the distinction between adaptationist and more developmentalist positions using the vocabulary of *externalism* and *internalism*. He gives this rough outline of the distinction: ". . . the term 'externalist' will be used for all explanations of properties of organic systems in terms of properties of their environments. Explanations of one set of properties in terms of other internal or intrinsic properties of the organic system will be called 'internalist'" (p. 30).

Whereas adaptationism is broadly externalist (ibid., p. 32), developmentalists are allied with internalism: "The properties of individual development from egg to adult, in particular, have long been thought to direct or constrain evolutionary change in this way . . . The language of 'constraint' in evolutionary biology is often the language of a moderate internalism, or of concession to internalist arguments" (ibid., pp. 37–38).

I have some misgivings about this distinction; explaining why I do will help the argument of the next section. Godfrey-Smith's intention for the distinction between internalism and externalism seems to be concerned with whether changes in morphology can be predicted primarily by attending to facts about environments or facts about development. It appears at first glance that there is a legitimate distinction to be drawn here, and one which overlaps nicely with adaptationist and developmentalist approaches. Godfrey-Smith points to the famous study by Kettlewell that we came across in the last chapter, and he suggests that it is a paradigm case of an externalist explanation: "There is no more celebrated case of selective explanation than the effect that air pollution had on the coloration of moths which are preyed upon by birds (Kettlewell 1973). The trees became darker, and the moths followed" (Godfrey-Smith 1996, p. 136).

If this kind of explanation, where organism follows environment, turns out to work for a majority of traits, then surely we show that external factors are more important than internal ones in shaping organismic form? Well, in fact we do not show that, because our recognition of what are and what are not selection pressures acting on a lineage is itself colored by what we think of as the internal constraints acting on the system, as the following example shows. A population of light colored moths suffers predation from birds, and resides on dark colored trees. A knowledge of the nature of the environment alone does not yield the prediction that the moths will change to match the color of the trees. There are many imaginable ways that the moths could change in response to the dark environment. They could secrete a chemical that alters the color of the trees to a lighter one. They could stop residing on trees, and start residing on some other surface to which they are better matched. They could secrete a chemical that interferes with their predators' visual systems. They could become poisonous to their predators. These are all ways in which moths could respond to the environment. If we predict that they do respond in a certain way—by changing their color, for example—that can only be because we implicitly add a range of assumptions about which phenotypes are most easily accessible, and this is itself information about the internal workings of the organism.

The question of which parts of the environment are open to alteration by an organism is one that is informed by constraints. If we brush off constraints as of little explanatory relevance, we might normally think that we arrive at an externalist position—externalist because evolution of the lineage is not internally constrained. Yet for environments to predict change it seems that we must keep some parts of them fixed, which is to say that we must accept the existence of some internal constraints that keep these environmental features immune from adaptive alteration by the evolving lineage. The surprising result of this line of thought is that internal constraints, far from being in tension with explanations of changes in form in terms of selective forces, are required if such selective forces, pressures, or problems are to be recognized. Without constraints, the environment itself would give no guide as to future adaptive trajectories.

Although I have been considering the distinction between internalism and externalism as one between whether internal and external factors are more significant in determining form, this does not do justice to the notion that many *contrasts* in form may be explicable in terms of internal factors rather then external factors. Whether the color of moth wings is best explained by internal constraint or external environment may depend on just what question we ask about those wings (see Lipton 1991 for further discussion of contrastive explanation).

If we ask why moths have dark wings rather than light wings we will give an answer in terms of selection pressures—dark-winged moths were better suited to the environment than light-winged ones. If, on the other hand, we ask why those same moths have dark wings rather than tree-changing chemicals, we will most likely give an answer in terms of internal constraint—the moth's history or developmental system does not make the secretion of tree-changing chemicals a possibility.

If the constraints working on diverse lineages are largely the same, then if we tend to ask questions that compare the morphologies of different lineages, the answers to those questions will have recourse to the different environments in which those lineages find themselves. A rather weak form of externalism (and an empirically implausible one) might assert simply that constraints are constant, so morphological differences between lineages are not to be explained in terms of them. Such a view would give explanatory importance to selection only by restricting the types of contrasts to be explained. At times, Godfrey-Smith himself gives a gloss on externalism that suggests just this position: "The position taken in this book is that being an externalist very often involves making an explanatory bet. Adaptationists bet that spending most of your time and energy investigating patterns of natural selection at the expense of genetic factors is better than spending more time and energy on the genetic system and less on patterns of selection" (1996, p. 53).

If many constraints are constant across species, then many differences between species will not be explicable in terms of different constraints; rather, they will be explicable in terms of different selective pressures.

This would hardly show, of course, that we do not need to know what these constraints are if we are to formulate hypotheses concerning just what selective pressures may have acted to produce differences among species. Unless we know what typical ranges of variants developmental systems make available, then we cannot formulate hypotheses concerning the differences in fitness of these variants in different environments. So this model of an "externalist" victory would still make the discovery of developmental constraints an urgent project. Showing that developmental constraints are constant would not show that they were not essential parts of a model explaining how traits evolve—they would simply be an invariant part of that model.

4.4.2 Problem Solving and Evolving

We can now answer a set of criticisms from Richard Lewontin of the conception of evolution as a response by species to problems posed by their environments. I will wait until chapter 6 before I explore the relationship between the concept of a selective problem outlined here, and the concept of a problem faced by a designer.

Lewontin argues that the conception of adaptations as solutions to problems is committed to a false picture of the relationship between an organism and its environment: "[Adaptation] is the concept that there exist certain 'problems' to be 'solved' by organisms . . . and that the actual forms of biological . . . organizations we see in the world are 'solutions' to these 'problems'" (1984, p. 236). His claim is that the problem solving conception is committed to a picture of lineages passively changing to fit a fixed environment. However, he argues that the relationship between organism and environment is really one of "construction": "[Organisms] are not the passive objects of external forces, but the creators and modulators of these forces. The metaphor of adaptation must therefore be replaced by one of construction, a metaphor that has implications for the form of evolutionary theory" (1985, p. 104).

Lewontin in fact invokes two quite distinct senses of "construction," as Godfrey-Smith (1996, section 5.4) shows. The first is a logical one, in which he argues that a niche cannot be defined except by reference to the organism that occupies it. The second is a causal one—organisms act so as to maintain and sometimes to alter the make up of their environments.

The second claim is obviously true, and the first one seems likely to be true also. I suggest that a suitable concept of "environmental problem" can be salvaged nonetheless.

Lewontin's first argument against the concept of adaptation draws on the logical sense of niche construction: "If organisms define their own niches, then all species are already adapted and evolution cannot be seen as the process of becoming adapted" (1984, p. 238). Perhaps it is true that organisms define their own niches. So let us suppose that the best way to pick out the rabbit niche is just to say that it is the set of environmental circumstances that affects the life of the rabbit. It would not follow from this that we could not pick out various adaptive or selective problems that act on the rabbit. We can equate the idea of a selective problem with the idea of a selection pressure outlined in chapter 2. A species faces the selective problem to F if and only if, were the species to acquire or increase its capacity to F, it would then become fitter.

Lewontin is right to say that an environment does not specify a template for ideal adaptation in the way that a lock specifies a template to which a key should ideally be matched. There is no "ideal species" dictated by the nature of a savannah environment, for example, and the question of what problems the savannah environment poses can only be answered if we first know what kind of species we are asking after—predator, prey, ground-dwelling, avian and so forth. So in this sense, Lewontin is right to attack the "lock and key" metaphor of adaptation.

Lewontin is also right to attack the implication of the "lock and key" metaphor that adaptation is achieved by conforming to an environment, rather than by altering it. We saw that we should not assume that moths need adapt by matching their background—they might also change it. The idea of a problem as a selection pressure makes room for this since it is true, for a group of moths residing on a dark background, that were they to alter their background to match their own color, they would become fitter. What is more, the idea of a problem sketched also makes room for the idea that by adapting, species can alter the nature of the problems they face. There is no contradiction in saying that in virtue of some selection pressure being partially met, the nature of the selective problems posed might be altered. The mathematical apparatus of evolutionary game theory gives us some of the tools we need to model situations where problems change as organisms adapt to answer them. A workable concept of an adaptive problem can remain intact even if it is true that organisms themselves are the creators and modulators of adaptive problems, just so long as the relationship between a lineage's response to problems and the consequent modification of those problems is not too chaotic.

So although I am sympathetic to Lewontin's insistence on the importance of construction, and the misleading nature of the lock and key metaphor, I am surprised that he takes his critique to undermine the concepts of adaptation and evolutionary problems, at least in the way that those concepts are used among adaptationists (again, I follow Godfrey-Smith 1996). Those biologists who are most enthusiastic about the existence of adaptation are often the keenest to acknowledge both that adaptive problems are not constant, and that organisms can respond to problems by manipulating, rather than conforming to, their environments. Evolutionary game theory contains plenty of examples of the first phenomenon, and the theory of the extended phenotype gives many examples of the second. Consider Mayr on the phenomenon of plants making galls to house insects. Mayr's talk of selection pressures is strongly adaptationist in style, yet it makes room for just the kind of construction highlighted by Lewontin:

> Why . . . should a plant make the gall such a perfect domicile for an insect that is its enemy? Actually we are dealing here with two selection pressures. On the one hand, selection works on a population of gall insects and favors those whose gall-inducing chemicals stimulate the production of galls giving maximum protection to the young larva. The opposing selection pressure on the plant is in most cases quite small because having a few galls will depress viability of the plant host only very slightly. (Mayr 1963, pp. 196–197; quoted in Dawkins 1982, p. 219)

Perhaps even more striking is the celebrated "brainworm" *Dicrocoelium dentriticum,* which is, in fact, a fluke. The fluke burrows into the suboesophagal ganglion of its intermediate host the ant, and causes the ant to climb to the top of grass stems, increasing the likelihood that the ant is eaten by an ungulate such as a sheep. Thus, by manipulating the behavior of the ant, the fluke continues its lifecycle (Dawkins 1982, p. 218; see also Wickler 1976 and Love 1980 for further detail).

One might fear that the account I have sketched of an evolutionary problem is too liberal, that it recognizes too many problems. If we allow, as has been suggested, that our moths face the problem both of changing to match the color of the leaves, and of changing the color of the leaves to match them, then we are committed to recognizing an infinity of outlandish selection pressures and problems. The problems faced by an evolving lineage do not include simply those alterations that are essential if it is to survive, but also those alterations that would increase its fitness. The totality of problems faced by a species is more like the range of fitness enhancing *opportunities* offered in an environment than the range of measures that are *necessary* for survival in that environment. The range of opportunities for fitness enhancement will be hard to restrict. Just as moths face the problem of altering the color of their leaves, so they face the problem of hypnotizing birds to ignore them, the problem of developing weaponry to destroy birds, and so forth. But, one might think, this range of problems must surely be limited.

In fact, I can see no good way of excluding these outlandish problems, nor do I think we need to. One might try to distinguish the two sets by saying that the only problems faced are ones that are solved. But this seems to be a corruption of the label "problem," in that a problem can surely exist without a solution being found, and it also robs the idea of a selective problem of its predictive power. The notion of constraint, a common one in evolutionary studies, is of some fact that prevents the production of an expected phenotype and hence, in some cases, the prevention of an anticipated adaptive pathway. Consider, by way of an example, these words from biologist Mary McKitrick: "Phylogenetic constraint . . . implies the absence of an anticipated course of evolution, such as, for example, the failure of birds to evolve viviparity [giving birth to live young]. I would define phylogenetic constraint, therefore, as any result or component of the phylogenetic history of a lineage that prevents an anticipated course of evolution in that lineage" (McKitrick 1993, p. 309).

If we are to leave room for constraint, we must allow that selective problems can go unanswered. So an appeal to constraint explains at the same time why we should be liberal in our recognition of adaptive problems, and also why the breadth of adaptive problems need not undermine the importance the problem solving approach has for the practice of biology. Only those problems that stand a chance of being solved to some degree will feature in predictions of likely change or explanations of actual change. The recognition of a selection pressure for armaments will not tell us of a likely future course of evolution among moths, nor will it help us to uncover a past course. Even the forms of constraint that have prevented moths from developing miniature machine guns will be so nebulous as to shed very little light on facts that might be of interest to biologists. We do not need to deny the existence of more outlandish adaptive problems to explain why one never hears mention of them in serious studies.

For the purposes of explanation and prediction, then, we must confine ourselves to those problems that stand a chance of being solved. This entails that we hold those traits that are unlikely to change constant—traits which, were they altered, would almost certainly have disastrous effects on viability or fitness—and also that we consider only those variants that we have reason to believe are permitted by the developmental system. My defense of the concept of an adaptive problem does not, therefore, show us that adaptationists should expect to be able to predict adaptive responses from broad knowledge of environment alone. This adds to my doubts from the end of chapter 3 about the uses of adaptive thinking in evolutionary psychology; in defending the concept of an adaptive problem I do not seek to downplay the formidable epistemic problems facing this discipline in its hopes to use such problems to uncover the workings of the mind.

My account of problem solving is similar to that advocated by Kitcher (1993). The main difference with Kitcher is in my understanding of a selection pressure: "In identifying the environment-centred perspective, I have explicitly responded to [Lewontin's argument], by proposing that the selection pressures on organisms arise only when we have held fixed important features of those organisms, features that specify limits on those parts of nature with which they causally interact" (p. 383). Kitcher's phrasing is curious; what we must limit in holding fixed certain features of organisms are the kinds of changes that an organism might realistically be expected to undergo. What is more, Kitcher's language suggests a confusion between the organism's niche in the sense outlined above—that set of facts that influence its survival and reproduction—and what might be called the organism's environment of adaptation—that set of facts about the environment which the organism is unable to alter over evolutionary time. Nevertheless, Kitcher's practical recommendations are similar to my own:

> Quite evidently, if we were to hold fixed properties that could easily be modified through mutation (or in development), we would obtain an inadequate picture of the organism's environment and, consequently, of the selection pressures to which it is subject. If, however, we start from those characteristics of an organism that would require large genetic changes to modify—as when we hold fixed the inability of rabbits to fight foxes—then our picture of the environment takes into account the evolutionary possibilities for the organism and offers a realistic view of the selection pressures imposed. (Ibid., p. 383)

Kitcher's position does not differ on any serious methodological issue—we both agree that any prediction of an evolutionary trajectory requires consideration both of the demands of the environment and of the possibilities afforded for change by the organism—however, where I speak of a multiplicity of selection pressures, with organismic possibilities determining which might be answered, Kitcher speaks of those organismic possibilities determining which selection pressures exist. I doubt if anything of substance turns on this difference.

5

Function, Selection, and Explanation

5.1 A Primer on Functions for Strangers to Swampman and the Pumping of Blood

As we have seen in previous chapters, biologists use words like "function," "purpose," and "design" to talk about organisms in the same ways that the rest of us use those words to talk about artifacts. The search by philosophers for an analysis of biological function often takes pains to show that this verbal isomorphism is justified, and it tries to do this by showing how biological properties and processes can support concepts with the same connotations as the concepts supported by the processes of artifact use and design.

The job for this chapter is to show that a variant on Cummins's (1975) account makes best sense of the practice of function attribution within biology. Note that it is one thing to give a clarificatory analysis of biological function, a different thing to explain why biologists, but not physicists or chemists, use teleological language. The second task is saved for chapter 6. Cummins tells us:

> x functions as a ϕ in s (or: the function of x in s is to ϕ) relative to an analytical account A of s's capacity to ψ just in case x is capable of ϕ-ing in s and A appropriately and adequately accounts for s's capacity to ψ by, in part, appealing to the capacity of x to ϕ in s. (p. 64)

In other words, once we fix some capacity (ψ) of a containing system (s) that we seek to explain, the function (ϕ) of any part (x) of that system is just the causal contribution it makes to that capacity. So the function of the heart in the context of the circulatory system is to pump blood; equally, the function of the heart in the context of hospital diagnostic practices is to make sounds of a certain type. Cummins's account—henceforth the causal role, or CR account—is liberal enough to license function ascriptions to nonadaptive biological items like tumors, and also to parts of purely physical systems like glaciers and clouds.

It is commonly thought that artifact function ascriptions have three connotations, and that biological processes can support function claims with these connotations, too. Later in this chapter I will cast some doubt over whether an analysis of biological function really does need to make sense of all three connotations, but for the moment they are useful in laying out the terms of the debate. For the reasons I give below, most philosophers reject the simple CR function concept because it cannot make good sense of these connotations.

1. *Function ascriptions are explanatory.* To ascribe a function is to give a teleological explanation. CR functions are involved in explanations, but they are usually the wrong kind of explanation to qualify as teleological. If we cite a CR function of the heart what we explain is something about how a heart currently contributes to a complex capacity. But an intuition most clearly articulated by Wright (1973) has it that teleological function claims tell us "why an item is there." This intuition is often backed up by examples of artifacts with functions that, as a matter of fact, they are unable to carry out. "The function of the curtain is to prevent people from looking in" tells us something about why the curtain is there; but it does not tell us how the curtain contributes to any complex capacity because, let us suppose, the curtain is full of holes and does not prevent people from looking in. If the function of an item can be an effect that item does not have, then the function of an item cannot be a causal contribution that item makes to anything.

2. *Function ascriptions make normative claims.* To say that the function of the heart is to pump blood is to say that this is what the heart ought to do, or that it is what the heart is supposed to do, or perhaps that this is what the heart has been designed to do. Biological function ascriptions thus make room for malfunction in a way that CR functions do not. If a heart cannot pump blood then it has gone wrong just as, if the curtain of our previous example does not frustrate voyeurs, then it has gone wrong. Perhaps the CR account of function can be augmented to accommodate normative function claims; however, for the moment it suffices to note

hat the basic CR idea of a contribution to a complex capacity is purely lescriptive.

3. *Functions can be distinguished from "accidents."* As a corollary of 2), not all of the effects of a trait, not even all of its beneficial effects, are unctions of the trait. A bible is for reading and for spreading God's word; a bible kept in a breast pocket that deflects an otherwise lethal bullet does not have preserving life as a function, even though this is a beneficial effect of the bible. This is merely an accidental benefit. Similarly, a kidney benefits the bearer by filtering blood; it might also benefit the bearer by fetching a good price on the transplant black market. But yielding money is only an accidental benefit of a kidney, because it is not something that the kidney is "supposed" to do. CR functions cannot distinguish functions from accidents: if the complex capacity to be explained is the cashflow within some household, then we may indeed have to say that the kidney of one of the householders makes a monetary contribution to this capacity, and hence has its transplant value as a function.

When we think about artifacts we might naturally suppose that the function of an item is just whatever it is intended to do. It is quite easy to see how an appeal to intention might satisfy our three connotations of function ascriptions. I place a jam jar in the middle of my floor, and a friend asks me what it is for. "To catch drips," I tell him. It seems that this claim reports my intention in placing the jam jar there, and it also gives a causal explanation in terms of this intention for the presence of the jam jar. The intention specifies what the jam jar is supposed to do, and thereby gives a clear criterion for when the jam jar is malfunctioning. Perhaps the jam jar was not rinsed thoroughly, and still contains a good deal of jam. As a result it attracts and traps in its sweet goo many troublesome wasps and bees. Attracting wasps is an accidental benefit, not a function of the jar; this is so because I have no intention that the jar should attract wasps.

All of this suggests the following naive account of function ascriptions for artifacts:

(IE) The function of artifact A is F iff some agent X intends that A perform F.

We can call this the intended effects (IE) account of artifact functions. The function of an artifact is just whatever effect its maker—or perhaps some subsequent user—intends it to have. I do not want to defend IE as the only plausible account of artifact teleology. IE would need modification (and see McLaughlin 2001, ch. 3 for suggestions) if we want to rule out the possibility that the sun has the function of playing compact discs merely in virtue of the fact that someone intends the sun to do this. It also seems likely that there are some function statements we make about artifacts that are not grounded in intentions at all (see Preston 1998). Yet for the sake of clarity in this chapter I will be comparing various accounts of biological function with the account of function put forward by IE. In chapters 6 and 7, where I discuss artifacts and artifact functions in more detail, I will argue that some of our function claims about artifacts do not follow this scheme.

Constructing a full-blooded function concept in biology now looks like a straightforward affair. Intending that an item do something, one might think, is simply to select it for some effect. So if the function of an artifact is what a designer or user intended it to do, then the function of a trait is what nature selected it for. This analysis will support our three connotations of function attribution in biology, with no need to posit a designer. Thus we arrive at the ubiquitous selected effects (SE) account of biological function that claims, roughly, that:

(SE) The function of a trait T is F iff T was selected for F.

This account has roots in Wright's (1973) etiological account of function. The account is called etiological because it is an aspect of an item's causal history—its etiology—that determines what its current function is. Just as the intentional account fixes the function of an item by reference to an intentional episode in the artifact's history, so the SE account of function fixes the function of an item by reference to its history of selection. SE accounts of this broad form are given by Griffiths (1993), Godfrey-Smith (1993, 1994), Neander (1991a,b), and Millikan (1989c) among others.

Selection, it is claimed, explains the presence of hearts by virtue of the effect of pumping blood. It is because ancestral hearts pumped blood that creatures with hearts had a fitness advantage over heartless competitors. Hearts thus increased their representation in the population of which their bearers were members, so this effect explains why there are hearts today. Because of this appeal to history, we also give sense to the idea that when a heart cannot pump blood it is malfunctioning. It fails to do that which caused past hearts to be selected. Finally, some of the heart's beneficial effects, although they may contribute to fitness or well-being now, are not effects that ever caused hearts to be selected. For this reason, producing diagnostic noises is excluded as a function and is instead credited as a mere "accidental" benefit.

The consensus position among students of biological function these days is one of pluralism (e.g., Godfrey-Smith 1993, 1994; Amundson and Lauder 1994). SE function is the concept invoked when many function claims are made in evolutionary biology. Moreover, the SE function concept expresses a quite genuinely *teleological* sense of function ascription. Yet not all biological functions are SE functions. In other branches of biology the CR function concept is involved, for example when one asks what is the function of free radicals in the formation of tumors. This is a wholly nonteleological function concept that corresponds to the causal role of an item in a system, and which lacks the connotations possessed by the function concept as it is used to describe the purposes of artifacts and their parts.

5.2 A Problem Case for the SE Account

SE accounts of function are too readily accepted, perhaps because their proponents overlook just what is involved in selection. I suspect that the move from an account of artifact function that refers to intention, to an account of biological function that refers to selection, is eased by the thought that intending that something do something is just to select it for those capacities. It is eased, in other words, by an appeal to the artifact model of evolution. Yet the strictly technical notion within biology of selection for some property is quite different to the intentional concept of selecting something under some set of criteria.

What does it mean to say that a trait is *selected for* some function? Sober coined the expression to mark a distinction between selection of objects and selection for properties. He offers the following summary:

> "Selection of" pertains to the effects of a selection process, whereas "selection for" describes its causes. To say that there is selection for a given property means that having that property causes success in survival and reproduction. But to say that a given sort of object was selected is merely to say that the result of the selection process was to increase the representation of that kind of object. (Sober 1984a, p. 100)

The distinction can be made clear using an example. The phenomenon of longshore drift results in an accumulation of small pebbles at one end of a beach, and larger pebbles at the other end. Suppose that the larger pebbles are made from a dark mineral, and that the smaller ones are made from a light mineral. The dark color of the pebbles does not cause them to accumulate at one end of the beach; their size does. Here there is selection for size, but only selection of colored pebbles. (Strictly speaking, there is no selection here because there is no heritability; stones do not reproduce. But the example suffices to illustrate the point about the causal aspect of the concept of "selection for.") In the case of Kettlewell's moths, dark moths are fitter than light moths in industrial areas because (it seems) dark moths are better camouflaged against trees. So there is selection for camouflaging in the moth population.

For selection to occur, there must be variation. A population of moths undergoes no selection, hence no selection for camouflaging, if there is no variation among them. The logical requirement of variation for selection has awkward consequences for any simple selected effects account of function, as the following thought experiment demonstrates.

Imagine a population of moths that reside for most of their time on green leaves. The moths are all bright orange in color, and stand out quite starkly against the green background. As a result of this, they are frequently eaten by birds. Suppose, now, that a large corporation builds a chemical plant nearby. The plant releases a thick orange smoke, which settles as a powder on the leaves around the moth habitat. The orange moths are now well camouflaged against their new orange background, and as a result, incidence of predation by birds decreases sharply. The moths now have far greater reproductive success; they live longer, and have more offspring. So the moths increase in fitness, and the moth population expands. There is, however, no variation in wing color and hence no selection of orange moths. So the orange wings do not have the SE function of camouflaging the moths in this scenario—not even several generations after the change of the leaf color from green to orange, when the moth population has exploded.

One might reasonably defend this result by appeal to the notion of *design*. That moth wings camouflage the moth is a matter of accident alone; it is simply a fluke that they happen to resemble the color of the polluted leaves. They have not been modified or "shaped" in any way to match the leaf color. The appeal to design cannot, however, be used to defend the SE account, because the SE account denies moth wings the camouflaging function on the grounds that there is no variation, not on the grounds that there is no modification to the wings. By introducing variation into our example—perhaps even one moth that is some color other than orange and which has less reproductive success than the orange moths—we are forced to say that the moth wings do have the SE function of camouflage after all.

Imagine, then, that the situation is slightly different. Our moth population lives on green leaves as before, but there are two equally fit variants to be found in the population. The vast majority of moths are orange, and a handful are red. The two variants are selectively neutral. At the moment that the leaf color turns to orange, the orange moths alone are camouflaged. Hence the orange moths begin to outreproduce the red moths in virtue of their better camouflage. In this situation there is variation, and the orange moths are selected for their camouflage. So the orange wings have camouflage as an SE function. The difference is represented diagrammatically in figures 5.1 and 5.2.

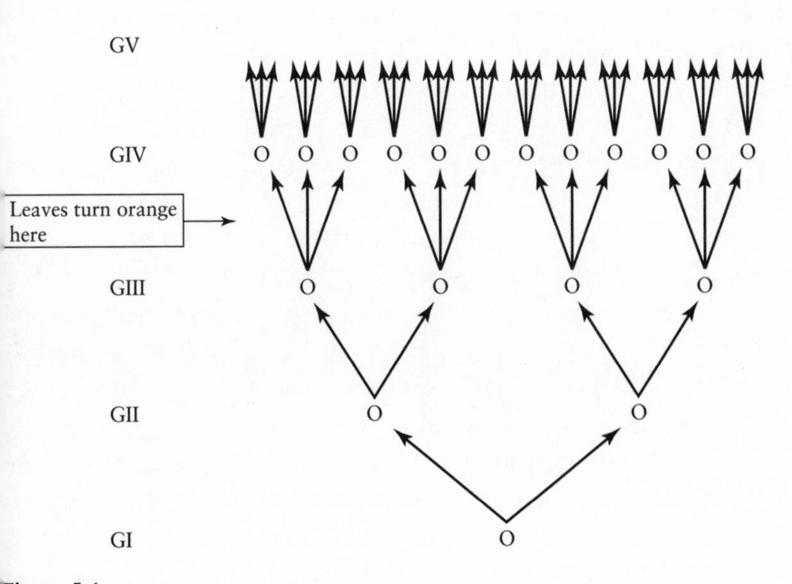

Figure 5.1
No variation, no selection, no SE function.

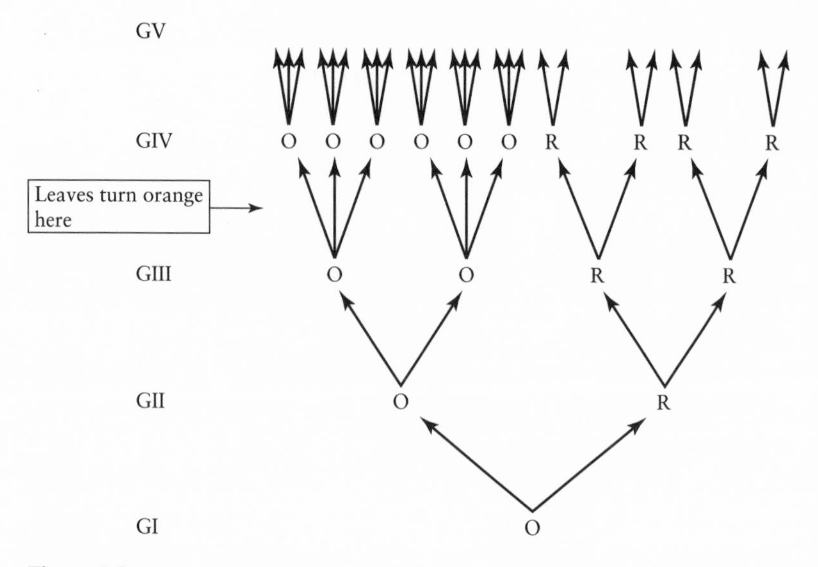

Figure 5.2
Variation, selection, SE function.

These scenarios help to illustrate a number of problems for simple SE accounts. As we saw earlier, a claim made in favor of SE accounts of function is that they do justice to the idea that function ascriptions explain why an item is present or "why it is there." If the SE account is to be justifiable as the *only* account of purposive biological function, one must show why the sense in which selection explains why orange wings are there in the scenario of figure 5.2 qualifies selection explanations alone for the role of functional explanations.

We can begin by noting one sense in which selection for camouflage does not explain why orange wings are there in figure 5.2: it does not explain the emergence of the orange wing trait type. Moths with identical orange wings existed in the population in figure 5.2 before there was selection for camouflage. This is why the notion of design cannot be used to justify the attribution of function in figure 5.2.

Perhaps selection explains why the trait is there in the different sense that it explains why there are so many orange moths in the population in generation five (GV). This suggestion is bad news for the SE account, because selection no longer seems necessary for function ascriptions. If the key sense for functional explanation is to explain why there are so many individuals with some trait, then whatever explains this fact in GV in figure 5.1 will also qualify as the basis for a functional explanation.

What does explain why there are so many orange moths at GV in figure 5.1? This depends on exactly which fact we are interested in. If we want to explain why there are more moths at GV than we would expect given the reproductive rates observed in GI to GIII, then we should cite the specific novel contribution to fitness of the orange wings that explains the increased reproductive rate that orange moths enjoy after GIII. If we want to explain the absolute number of orange-winged moths in GV, then we should maybe cite every contribution to fitness that the orange wings make, or maybe we should cite all the contributing factors to fitness of every trait of the organism. Whichever option we choose, it seems that we will be able to explain the proliferation of moths over several generations not by reference to selection, but by reference to some heritable contribution to fitness. So heritable contributions to fitness can explain, in a variety of senses, why the trait is there. The SE account's denial of the camouflaging function to moth wings in figure 5.1 seems arbitrary in the light of the intuitions that motivate it.

The SE account also entails that traits that go to fixation under drift should not be given functions; however, when these traits are selectively neutral, biological practice is divided. Kimura, for example (1991, p. 3) is clear that when he speaks of selectively neutral traits he does not wish to imply that they are function*less*: "... the neutral theory claims that the overwhelming majority of changes at the molecular level ... are the result of random fixation of selectively neutral mutants through sampling drift in finite populations ... Here "selectively neutral" means selectively equivalent ... In other words, mutant forms can do the job equally well in terms of survival and reproduction ... of individuals" (ibid.). Kimura understands neutral traits to be traits that perform the same function equally well—they are not traits without functions.

The basic motivating idea behind the SE account—namely that having a function is a matter of having a past effect that explains current presence—tells us that we should attribute functions to some traits that drift to fixation. If two variants are equally well camouflaged, and if one type drifts to fixation, then the camouflaging effect explains why the trait reaches fixation in the following sense: had the trait not had this effect, then it would most likely have been selected against. So this is another case that strains the SE account.

5.3 A Digression on Teleosemantics

It is worth pointing out that our problem cases have implications for "teleosemantics" (see Papineau 1993, and Millikan 1984, 1989b for examples of such theories). Teleosemantics makes the content of a belief dependent on its biological function, where biological function is typically spelled out using one of the standard SE analyses available. Opponents of teleosemantics are fond of invoking Swampman—the atom-for-atom replica of a human being, who has no ancestors and no evolutionary history and coalesces one day from swamp gunk. Swampman has no history of selection, hence no biological functions, and hence no content to his beliefs according to most teleosemantic accounts. However, say Swampman's champions, he is identical with a human who does have this kind of evolutionary past—he can presumably do all the same things as his doppelganger. So why is he denied content to his beliefs?

What the examples make clear is that it is not only the direct ancestry of a trait that matters for the possession of a given SE function. Suppose an organism bears some trait that contributes to fitness, and that is inherited from a long line of ancestral traits that made an identical contribution. If those ancestors were not accompanied in the population by others who lacked the trait in question, then whatever contribution ancestral copies of the trait made to fitness, the descendant still does not have that contribution as an SE function. If a teleosemantic theory is based on the SE account, then Swampman needs not only the right kinds of parents and grandparents for his beliefs to be contentful; he needs the right kinds of aunts and uncles, too. Those who find teleosemantics hard to swallow should find versions based on SE accounts especially troubling.

Mental states such as beliefs are commonly held to be normative states. This is on the grounds that beliefs can fail to represent their objects properly; that is, they can be false. When they do this they go wrong. Many teleosemanticists have hoped to make sense of this normativity by appealing to natural selection. Later in this chapter I will suggest that the selected effects account is bolstered by an improper analogy between natural selection and intentional selection. But the analogy with intentional selection also helps to create this impression that natural selection could have a distinctive role in creating normative states.

When one literally chooses, or selects, an item for a purpose, then one does indeed impose a norm of performance on that item. One thereby acquires an expectation for how the item should perform. Turning to our moths in figures 5.1 and 5.2, it is hard to credit selection with any distinctive role in imposing norms on our moth wings. In both cases they have historical norms to live up to; in both cases, that is, there are past tokens that have contributed to fitness in a characteristic way. So selection is not necessary for the imposition of historical norms in this sense. There are wings in figure 5.2 that have effects that other wings do not have, but that is true in figure 5.1, too, where earlier wings do not offer camouflage. So the mere existence of variation in performance is also not enough to single out the norms grounded by the SE account. There are no synchronic variations in performance in figure 5.1, but it is hard to see why that should be the important source of normativity to which the SE account lays claim, and in any case, there are rival accounts of function to the SE account (for example, Boorse's 1976, 2002 account) that also yield such synchronic norms by appealing to typical performance by tokens of some trait type at some time. In brief, it is hard to see what important norms are at work in figure 5.2 where selection acts that are not at work in figure 5.1 where there is no selection.

5.4 A Weak Etiological Alternative

A perceived strength of SE accounts of function is, as we have noted, their satisfaction of the intuition that to ascribe a function to a trait is to give a historical explanation of why the trait in question is there. CR function ascriptions merely make a claim about how a trait contributes to a complex capacity. However, if the complex capacity in question is reproduction, then by contributing to reproductive success, traits also cause their own representation in future generations. Buller defends an alternative etiological theory of functions that draws on this fact. His "weak theory" (henceforth WT) tells us that: "A current token of a trait T in an organism O has the function of producing an effect of type E just in case past tokens of T contributed to the fitness of O's ancestors by producing E, and thereby causally contributed to the reproduction of T in O's lineage" (1998, p. 507).

Thus WT has it that the moth wings in figure 5.1, at least in generations GIV and GV, have the new function of camouflage, because earlier wings contributed to the fitness of moths by camouflaging them. WT also gives functions to traits that are present because of drift, on occasions where those traits provide a fitness benefit.

There is, however, a problem with WT that repays investigation. By explaining how an organism is able to avoid predators and eventually reproduce, a trait like wing color partially explains why moths in their entirety "are there" in the next generation. The camouflaging effects of moths' orange wings explain the presence in the next generation not only of more orange wings, but of whole moths and all of the parts of the moth. They explain why the moth's legs are there, and its antennae, as well as its wings. Equally, the current presence of moth wings is explained just as much by the fitness contributions of ancestral moth legs and moth antennae, as it is by ancestral fitness effects of moth wings. This means that WT functions do not meet the requirement that a function ascription should explain why the functionally characterized item (rather than the whole organism) is present all that well.

It might be thought that this gives the SE account an advantage over WT. The camouflaging effect of orange moths in figure 5.2 explains the presence of later moth legs and later moth wings. However, the camouflaging effect of orange moths here explains why the orange moth type increases in frequency without explaining anything about the frequency of the legged moth type. That is so because the effects of orange wings control the frequency of orange wings without controlling the frequency of a universal trait like legs.

This specificity of functional explanation seems an important desideratum because we feel that biological processes should license the kind of specificity that the IE account enables for artifacts. Our assumption was that when we give the function of some artifact (or part of an artifact), we describe the intention of an artificer or user for what effect that artifact, or part thereof, should have. Suppose, then, that I add a rack to the roof of my car with the intention that it should carry luggage.

This explains the presence of the luggage rack as a whole, and it does not explain the presence of other parts of the car. Because the effects of ears, on the other hand, can only explain the presence of later ears by explaining the presence of other parts of the body, the effects of ears on fitness cannot specifically explain the presence of later ears in the way that the intentions for some part of an artifact can.

It turns out that not even SE functions always achieve the degree of explanatory specificity that we might want of a function concept. Consider a case where traits are linked, so that having orange wings is developmentally bound up with having hairy legs, say. Imagine that the orange-winged moths in figure 5.2 have hairy legs; the red ones do not. In this case the contribution of assisting in camouflage explains both why the orange wing trait increases in frequency and why the trait of having hairy legs increases in frequency.

These kinds of cases (the phenomenon is called "pleiotropy") where traits are linked are surely very common in organisms. They have an important impact on the SE account, in the light of its basis in the intuition that functions explain why a trait is present. If we choose to think of the function of a trait as whatever ancestral effects explain why so many organisms of that type are present over some other type, then we will be forced to say in this case that a function of the hairy legs of moths is to assist in camouflage. Assisting in camouflage is an ancestral effect that explains why moths with hairy legs are present over alternative non-hairy-legged moths. Of course, assisting in camouflage is not an ancestral effect of *hairy legs* that explains why the hairy-legged type is there, but then neither is assisting in camouflage an ancestral effect of moth legs that explains why they are there in the explanatory sense of WT. Neither SE nor WT truly say that the function of a trait is just any effect that explains why the trait is there.

What this shows, significantly, is that neither WT nor the SE account suffices as a theory for the individuation of trait types. It is not always clear whether that is a goal for theories of function. Karen Neander (1991a, p. 180) seems to imply that it is, when she tells us that "most biological categories are only definable in functional terms." What should now be clear is that the SE account presupposes an account of trait types that cannot itself be functional. The SE account tells us that the function of a trait is the effect of previous tokens of traits *of the same type* that explain the selective success of that type. Only by adding this restriction do we avoid the consequence that the SE function of hairy moth legs is to assist in camouflage.

The obvious place to look to individuate trait types is in the concept of homology. So the SE account, and WT, should both determine the function of a trait by the typical contribution to fitness made by earlier homologous traits. Davies (2000a,b, 2001) believes this commits both the SE theory and WT to some form of circularity. It is not clear to me why. Take a case where we wish to say that the function of the heart is to pump blood. For both the SE account and WT, this is grounded in part by the fact of earlier traits of the same type pumping blood. Here "earlier traits of the same type" means earlier homologous traits. Davies's worry seems to come from the fact that among earlier homologues, many of them may not have been able to pump blood. That is, many ancestral hearts may have been valveless. However, so long as we can individuate trait types by reference to homology, then we can ask why traits of that type spread, or proliferated, even when some traits of that type made a different contribution to fitness, or indeed no contribution at all. So long as some of the ancestral traits had a fitness enhancing effect, then there is an effect of earlier tokens of the type that explains the spread of that type. Homology identifies trait types, while the typical contribution to fitness that explains the selective success or reproduction of some type tells us what its function is.

Finally, I turn to an argument raised by Millikan against WT that claims that the concept of a trait's contribution to fitness is meaningless unless that trait is contrasted with some other trait that makes a different contribution. So, she says, only the SE account can offer a theory of functions, for only in the context of competing traits can we say what the contribution of any one trait type is to fitness:

> A trait that enhances or would enhance fitness is one such that, on average over the actual individuals in the species, having it would produce a more fit individual than not having it. There is a reference here to counterfactuals, to what the fitness values of various individuals that have the trait would have been if they hadn't had it, to what the fitness values of various individuals that don't have the trait would be if they did have it ... But exactly in this sort of context counterfactuals are most notoriously indeterminate in truth value ... The notion of superior fitness,

as actually used in evolutionary biology, is a well-defined notion only because it is never taken to attach to any trait in a vacuum or absolutely, but only relative to alternative traits actually found in the population. (Millikan 1989a, p. 174)

I have two comments about Millikan's argument. First, even if we grant that we can only talk of a contribution to fitness made by a trait if we can contrast that trait with some actual trait that has a different effect, then we can still assign meaningful fitness contributions to traits where there is no selection. That is certainly the case in figure 5.1 above. The orange wings make a novel contribution to fitness, and we can see this by contrasting their effects not with those of competing traits (there are none) but with the effects of earlier tokens of the orange wings themselves.

Second, Millikan's argument seems to me too strong for the problem at hand—at least if we construe it as a claim about the metaphysics of causation. If it works, then it works against all attempts to give causal analyses of the workings of complex systems, and indeed all causal analyses. A boy throws a brick at a window. Did the brick cause the window to break? It seems to me that if we read Millikan's argument metaphysically, then there is no fact of the matter unless there have been a series of similar, contemporaneous, throwing incidents where bricks are thrown in different ways, or where they are thrown and miss, and where the window does not break, or breaks in a different way. We would be unable to make claims about the effects of parts of unique systems or processes—the Big Bang, for example—since there are no alternative processes to compare them with.

Millikan's claim looks more plausible as one about the epistemology of causation; that is, about how we find out whether an event has a certain effect. If we are to tell what effect the throwing of the brick had, we maybe need to compare the brick throwing with other throwing incidents. But that can be achieved by comparison with our past experience of throwing incidents that occurred at different times and places, or even by setting up experimental situations where we alter the parameters of throwing. Equally, we can tell what contribution a moth's wing color makes to fitness by devising an experimental setup where we paint the wings of moths different colors and observe the effects in the wild. We do not need to assume that there was variation in the natural population itself, even if we grant Millikan the epistemological claim.

WT looks to be just as good a theory of functions as the SE account. It is no worse off than the SE account when it comes to explaining "why the trait is there." So, the first connotation of teleological function attributions is met. The second and third connotations are met also. WT can distinguish functions from accidents; accidental benefits are just fitness benefits that a trait's ancestors did not have. And WT makes room for a sense of "malfunction" in the same way that the SE account does. A trait is malfunctioning on the weak account just in case it does not have the capacity which ancestral functional tokens of the same trait type had. WT has all the strength of the SE account, and perhaps does better on some cases of function ascription.

5.5 Nonhistorical Functions

We just saw how an account of functions can do away with an appeal to selection, yet still meet the criteria for purposive function attribution. I now want to show why an account of functions can also do away with an appeal to history by thinking of functions not as past fitness contributions, but as current fitness contributions. First, I want to outline this naive fitness account of function (NF). I do not think that NF is the only defensible analysis of a teleological function concept that applies to biological systems. But I do think that NF comes closest to what evolutionary biologists should want from a function concept.

The naive fitness account says:

(NF) The function of a trait *t* is *F* iff traits of type *T*, of which *t* is a token, make a significant contribution to fitness by performing *F*.

As with SE and WT, trait types must be taken to be homology types. NF is a nonhistorical theory similar to Bigelow and Pargetter's (1987) propensity account, according to which functions are simply propensities of traits to be selected. NF is more liberal than the propensity account, however, because NF accords a function to a trait when it makes a contribution to fitness whether this contribution disposes the trait to be selected or not. A trait is not disposed to be selected when there is no variation, even though it may make a fitness contribution.

NF is able to satisfy at least the second two connotations of function ascription. Functions are distinguished from accidents: accidental fitness contributions will be those that arise from freak developmental or environmental circumstances and hence are manifested in only a handful of traits of the type under investigation. A trait will be said to be malfunctioning just when it fails to have the fitness contribution made by other traits of its type. I shall come back to how NF fares with regard to the first (explanatory) connotation in section 5.7.

Nonhistorical theories of function have been attacked from many quarters. Most objections focus on the problem of determining which environment should be used to evaluate the contribution a trait makes to fitness. It is obvious that fitness contributions vary with environment: orange wings provide camouflage against orange backgrounds but not against green backgrounds. The answer to the problem of fixing environment is quite simple: we should assess functions over the environment in which the trait in question typically finds itself. NF allows that functions change over time. The function of a trait at a time *t* is simply the contribution to fitness the trait makes in its typical environment at *t*.

Still the account remains vague. Two questions in particular demand answers. First, what of a trait that makes some contribution to fitness, but where that contribution is only quite weak? In other words, how great does a fitness contribution need to be to count as a function? I do not think this vagueness should be too worrying. The SE account itself faces similar problems: how significant must an ancestral fitness contribution be for it to count as a selected effect? If this kind of question can be written off as natural looseness in scientific terms, then we can say the same for the looseness in the functions attributed by NF.

The second question is more threatening. What if 5 percent of traits have some effect on fitness, while 95 percent do not? Is the effect then an "accident"? Or should we instead credit it as a function, and say that the remaining 95 percent are malfunctioning? In short, what proportion of the traits in the organism in question need perform *F* for the remainder of traits of the same type to have *F* as a function?

It is true that similar difficulties face the SE account. There is no easy answer to the question of just how many tokens of the type need to have had the effect *F* for it to count as selected. Even so, NF threatens to give completely the wrong answer in cases where what we most naturally wish to call a function of some trait is, in fact, a wholly atypical effect.

An example of this kind of case is given by Sterelny et al. (1996). Most acorns rot, rather than growing to produce oak trees. So growing is not a typical effect of acorns. Yet it seems important that biologists should be able to credit acorns with the function of producing adult trees. What we need, then, is a way of reading the claim that trait *t* has a significant effect on fitness in virtue of effect *F* that does not reduce to a statistical claim, yet which also does not turn NF into a version of the SE account or WT.

The way to do this is to remember that, for evolutionary biologists, the presence of traits in a population is the fact of primary explanatory interest. Hence the kinds of effects that are significant are those that can maintain a trait in a population, or which can increase the absolute number of some trait in a population. Since the effect of acorns on growth of trees is strong enough to maintain the presence of acorns in the population, it counts as a function in spite of its relatively low frequency in the population.

Does this concession reduce NF to a version of an etiological theory? No: NF gives functions to classes of homologous traits according to their effects at the time in question, not according to their past effects. NF functions cannot, therefore, explain the presence of the very items that are functionally characterized; for nothing, not even a set of traits, can explain its own presence. NF can elude the case of the acorns without collapsing into a historical theory.

NF allows functions of traits to vary according to how broadly the homology type in question is understood. That can be dictated by explanatory interests in a way that seems attractive to me. So we might class some group of enzymes as homologous across all insects, say, and ask what their function is in insects. Some of those same enzymes might be given more tightly circumscribed functions when we think of their functions in mosquitoes, or in some particular subpopulation of mosquitoes. In each case the tokens united under a homology type, and also the typical environment of those traits, will vary, and the attributed functions will vary with them.

Is not homology itself a historical concept? Does this turn NF into an etiological theory? Two short comments suffice here. First, not all definitions of homology are historical (see e.g., Jardine 1967; Jardine and Sibson

1971). In some cases, the concept of homology is defined by a measure of similarity in the relative position of a trait between two species. It is true that even this type of theory may impose a mild historical condition such that relative position is determined in part by the positions of the traits at the embryonic stage. This is an etiological definition of sorts, but a weak one where etiology stops short at the developmental history of the organism, not its evolutionary history.

Depending, then, on just how we understand homology, it is open to NF to attribute functions even when no relations of descent hold between the traits under investigation. In other words, the "instant lions" that are sometimes imagined to coalesce from nothing, unrelated to each other, may have NF functions. Even if we choose an evolutionary definition of homology, so that NF is shown to have strong historical commitments in its appeal to homology, this would not show that it collapses into an account of function that is the same as the etiological accounts considered above.

Some argue that the idea of a normal or typical environment for a trait should only be understood historically. So, were we to transport a group of mice, say, to a new environment where their large ears camouflage them against predators in the earlike grass, surely we would not want to say that their ears have camouflage as a function? That is just the kind of intuition exchange I have tried to avoid in this book. Of course in the mouse case the novel contribution to fitness is not something ears are designed for, but we have seen already that simple SE and weak etiological accounts also fail to capture this idea of design. Still, I owe some kind of account for why the immediate acquisition of novel functions fits with biological research, and why we should not worry that NF may offend some of our intuitions. We can give this kind of account by examining an argument from Karen Neander.

Neander (1991a, p. 182) argues that accounts like NF will give the wrong result when we encounter actual environments that depart from their historical precedents. Take a case where kidneys have filtered blood in the past, but where a sudden worldwide administration of some toxin means that no current kidney is able to filter blood. Neander wants to say that all kidneys are malfunctioning here, but for NF we need to say instead that the kidneys have no function. So, says Neander, accounts like NF must be altered to allow us to say that all instances of a trait are malfunctioning at some time *t*.

Our problem arises from the vagueness in the extension of "current environment" of the kidney. It is clear enough that "current environment" should not be taken to mean some infinitesimally small time slice of an environment that might correspond to "the present." For traits to discharge their functions at all—for them to make contributions to fitness that are significant in the context of their population—some fairly extended time slice of an environment is required. The etiological theorist will retort that "current environment" will be extended backwards to include recently past environments, and in making such an extension NF becomes a variant on WT.

In response, we do not need to think that "current environment" picks out times only a little before the moment of utterance. When we speak of past functions, this problem disappears. It is obvious that in asking "What is the function of the sail on *Dimetrodon*'s back?" we are speaking about an extended time period—not just one generation, or a few years in a generation. If we thought that for one generation within this period all *Dimetrodons* had sails that failed to regulate heat, then our assessment that they were malfunctioning at that time depends as much on their atypicality with respect to (then) future *Dimetrodons* as (then) past ones. Because the failure of the sails to regulate heat is short-lived, we can recognize it as a failure to make the contribution to fitness that *Dimetrodon* sails of that time period make.

When we are thinking of functions now, we should also think of "current environment" as extending both backward and forward in time from the moment of utterance. Just how far forward and how far back will the current environment of the trait extend? That will most likely be dictated by the questions asked of the traits under investigation, and the generation time of the organism under investigation. Unless there is a fairly well-defined context of utterance, a question like "What is the function of kidneys now?" cannot pick out any time period unambiguously. Since the concept we are considering is specifically a biological function concept, and since we are considering the context of evolutionary biology, it is fairly clear that the typical time period of interest will extend for longer than a generation, and hence, in the human context, longer than a year or so.

A evolutionary biologist simply would not ask "What is the function of human kidneys now?" where "now" picks out the year 2003 alone.

Let us return now to Neander's thought experiment. What if a toxin is administered today, all kidneys lose their blood filtering powers, yet the inability of kidneys to filter blood will be short-lived? Let us suppose that the toxin will leave the water after only a few years, and kidneys will return to their filtering ways. Here there is no need to think of kidneys as losing their function at all, since the collection of kidneys for which we assess typical fitness contribution comprises many future kidneys that do make the contribution of filtering blood. Since the transience of this loss of effect will in many cases be clear to investigating scientists, the extension of the class of traits into the future need not present insurmountable epistemic burdens to a biologist who wishes to make some claim about function or malfunction. If, on the other hand, the toxin is present over a significant time period, then we will have to say that kidneys lose their function when the toxin is present. Perhaps it is true that this offends our intuitions when we think of artifact functions.

Maybe some of us will be inclined to think instead of all the kidneys as malfunctioning across this period. Still, this kind of intuition should not make a difference when our goal is to give an account of a function concept suitable for biologists' purposes. That point is made clear when we remember that, in the terms of this thought experiment, the toxin becomes part of the normal developmental environment for kidneys. Kidneys incapable of filtering blood are the expected universal outcome in the changed environment. In that respect, kidneys that are unable to filter blood have the same developmental status as a trait that acquires a new contribution to fitness owing to a new persistent feature of the developmental environment. Consider a case where, because of the presence of some new mineral in an environment, moth wing development alters so that the wings now match their background in color. The idea of the moths acquiring a novel function in virtue of this persistent new fitness contribution does not seem so hard to swallow, from the perspective of biological investigation. Neither should the idea that kidneys might lose a function in virtue of a reliable, new environmental feature seem strange from that same perspective. Of course *we*—as humans rather than as biologists—might want to say our kidneys are malfunctioning over an extended period in which they fail to have their fitness effect. That, however, is because they also persist in failing to provide a benefit that we desire, and that we have grown to expect.

Now we have three function concepts—SE, WT, and NF—all of which meet some of the connotations of teleological function claims, and none of which have inconsistencies. How should we choose between them? In the next section I argue that we should not expect any fact to point decisively to one of these accounts as the correct account of biological teleology.

5.6 There Is No "Killer Intuition"

What should we say to someone who wants to defend the SE account against WT, or against NF? I have demonstrated no inconsistency in the SE account, and I have tried to show that measured against some of the intuitions of what a theory of functional explanation should achieve, WT and NF have some advantages over it. But this hardly shows that SE is bankrupt.

The problem in making a decisive choice between the theories is that there is really no single "killer intuition" that will tell us which of the accounts is right. We have seen in previous chapters that artifact talk gets into modern evolutionary biology through application of the artifact model. It is through heuristic reasoning about likely responses of lineages to their environments, and about the ways in which complex adaptations can emerge through the action of selective forces, that biologists come to use terms like "function," "purpose," and "design." The problem in choosing a single analysis of function is that there is more than one way to tighten these terms in order to correspond formally to specific biological processes or properties, and none of them perfectly matches the connotations of teleological language as it is used in the context of describing and explaining artifacts. Some philosophers (most recently McLaughlin 2001, p. 61) advocate ignoring intuitions relating to artifacts when we come to construct, or elucidate, a biological function concept. But attention to artifact teleology is important in the context of these projects because, in noting the parallels between the ways in which we use language in the two domains, and the differences between processes in the two domains, we arrive at a principled explanation of the impossibility of any clean resolution to the debate over how to analyze biological functions. That does not mean that we cannot choose between analyses of function on the more pragmatic grounds of how well they meet the needs of biological practice, and I will argue in the next section that such a pragmatic analysis should lead us to favor NF.

I will begin this line of argument by contrasting WT and the SE account. The problem in choosing between them comes from the concepts of problem and choice themselves. We tend to think that an artifact has a function just when it is used for some effect, or when the artifact is thought to solve some problem. Malfunction claims then arise when the artifact does not have the intended effect, or when it cannot solve the intended problem. Both the weak theory and the SE theory attempt to transfer these intuitions for what counts as an artifact function to the biological realm, yet neither theory can do this perfectly. That is why there is no way that comparison with artifact cases can tell us unequivocally that one account is right and the other wrong. Indeed, in other theoretical and explanatory contexts—developmental biology, or the social sciences, say—we should expect the possibility of the construction of yet more function concepts, which may also meet the connotations of artifact function attributions with varying degrees of success.

The SE theory derives some of its strength as an account of function attribution from its superficial congruence with the idea that artifacts get functions when they are selected for certain effects. If a rock has the function of weighing down papers on my desk, then we might think that is so in virtue of my having selected the rock because I think it will be good at weighing down papers. On the face of it, Sober's idea of "selection for" in biology mirrors this perfectly. Just as an agent can select a rock for some of its capacities, so an environment can select among a collection of different moths for their capacities (Wright 1973, 1976 stresses these kind of analogies). But there are really two quite different concepts of selection at work here. The intentional concept is that of picking out some item in virtue of some perceived capacity it has. The concept of natural selection is of one type outcompeting another in virtue of some causal capacity.

There are three important differences between natural and intentional selection. First, natural selection relies on the existence of different competing types, whereas intentional selection does not. A rock can be selected for use as a paperweight without it being compared with other items that have different capacities. Second, natural selection involves at least one of the competing types having the selected capacity, while an agent can select an item on the basis of a merely perceived capacity. A rock can be selected for use as a paperweight even when it is too small or too light to hold down papers. Third, as we saw in chapter 2, natural selection is essentially a population-level phenomenon, while intentional selection is not. An individual can reach out and select a single item in virtue of some capacities. Although selective forces act on individual organisms, natural selection is not a force that acts on individual organisms in this way. That is why, to repeat a point made in the previous chapter, it is possible for some population of artifacts to undergo no population-level selection even when the individual artifacts in the population are intentionally chosen by their users for certain capacities. If car buyers are just as likely to prefer red cars as blue, then red and blue can undergo no population-level selection even though each individual car is intentionally selected for its color.

McLaughlin (2001, ch. 7) also stresses many differences between natural and intentional selection, but the two of us differ on the moral to draw from them. McLaughlin (p. 150) thinks the disanalogies show that any attempt to use the artifact model as a key to understanding teleological language in biology is misconceived. My view is that attention to the artifact model and to the strong disanalogies between intentional and natural selection are crucial to explaining how artifact talk gets into biology, and why conflicts over the proper analysis of this talk cannot be decisively resolved.

We have a choice of how to translate the idea of intentional selection into the biological realm. Natural selection does not fit with intentional selection as well as its name might indicate. We might decide that part of the idea of intentionally selecting an item for its effects is that the item passes some kind of test, or matches some kind of criterion, laid down by the user. That does not entail the existence of competitors—an item can meet some set of demands on its own, without others having to fail. So we might now think that the weak theory offers a better analogue than the SE account to the concept of intentional selection. In the case of the moths, we wish to say that the environment lays down certain demands—certain possibilities for increasing fitness—which, when answered, should count as functions whether those demands are met by outcompeting alternative traits or not. So in this sense we say that the new orange tinged environment in figure 5.1 establishes the attainment of orange camouflage as a problem, and that orange-winged moths exist in later generations because their orange wing pigments have met that problem.

This is not perfect either. We might want to say that an artifact malfunctions when it cannot solve some problem that its designer lays down. Transferring this intuition to the biological case, we will say that an organism malfunctions when it cannot solve some environmental problem posed. Yet we saw in chapter 4 that the idea of an environmental problem is hard to restrict in its extension. Consider again the example from Symons (1992), where he suggests that there is an environmental problem facing human males to detect females who are fertile. Suppose we agree that were males to have this ability, they would be fitter. It is a problem, but a problem that is not solved, let us suppose, owing to constraint. On this kind of view we would have to say that because the human lineage has not been able to answer the problem, it is malfunctioning. We would then need to say in turn that, since rabbits fail to address the problem of evolving machine guns to ward off foxes, they are malfunctioning in this respect also. As I argued in chapter 4, this is not a practical difficulty for biology; however the SE account seems to do better than the weak theory on this score. For artifacts, the range of problems that an item should solve can be restricted to the recognized desiderata of inventors and users. SE is also able to restrict the range of functions that an item should discharge by reference to what the item has in fact been selected for, rather than the entire range of selective demands imposed by the environment.

NF is not a historical theory, hence NF does not tell us that functional traits are those that are present because they answer environmental problems. So one might think that NF is weaker than either etiological account. Unless NF explains "why a trait is there," then even if it tells us how the biological function concept should be understood, it does not count as a teleological theory of functions.

I have two broad lines of response to this argument. First, it is worth pointing out that what IE functions explain is often quite different to what etiological biological function concepts explain. When we see this, we see that the claims of etiological analyses to provide the most "genuine" teleological function concepts are strained, at least if we assess that claim by reference to the IE concept. There is a great difference in the kind of explanation offered by saying "He put the chair there because he thought it would help him reach the light bulb," and "Hearts are there in part because they have contributed to the reproductive capacities of their bearers." In both cases we might say that we explain "why the item is there," but this blanket locution hides the fact that the explanatory focus is very different in each case. In the first case, when we attribute a function to an artifact, what we really do is attribute a set of beliefs and goals to an agent who comes into contact with that artifact. The capacities of the artifact itself—past and present—need not feature. Yet in the biological case the capacities of earlier items of the same type as the biological item are central. At the limit, objects could have IE functions where their presence is not truly explained by the function attribution at all. A large, flat, immovable rock, used as an altar, certainly has an IE function, but the function attribution does not explain the presence of the rock—the rock would have been there regardless of the actions of those who use it. Reflection on our attribution of functions to artifacts casts doubt on the idea that the historical explanatory connotation is always a feature of teleological function attributions.

Second, although it is true that NF function claims do not offer explicit explanations of the presence of the traits to which functions are ascribed, they can often contain implicit explanations of presence. Current contributions to fitness by traits cannot explain the presence of those very traits. However, since NF functions will nearly always name contributions to fitness that are also the contributions of earlier traits, we can see that when we learn an NF function, we also usually learn something about the presence of the trait. NF functions offer implicit explanations—that is, they tell us what we need to know to be able to reliably infer an explanation in most cases—even if they do not explain presence explicitly.

5.7 A Pragmatic Defense of NF

How do the various accounts offered sit with respect to the account given in the previous chapter of reverse-engineering and adaptive thinking? One might think that the artifact model can only work in the context of an SE account, yet that is wrong. The artifact model, and the practices of explanation and prediction that accompany it, are just as well suited to WT and NF.

Imagine a natural population of light green moths residing on dark green leaves. We ask "What is the function of the green coloration?" and we answer "To give camouflage." Now the reverse-engineer or optimality modeler may be troubled by this. If it is true that camouflage is the function of moth wings, then we might have expected to observe a better matched, darker green moth instead. That kind of moth would have answered the problem of camouflage better. That thought prompts a number of ways to save the hypothesis of moth wing function. Perhaps providing camouflage is not the only function of moth wings. Or perhaps there was some constraint so that no darker variant existed. In this case we do not need to posit additional functions. Or perhaps drift prevented the darker variant from going to fixation. If we think that a function claim can be saved in this way by pointing to drift as an explanation for why the predicted better variant is not present, then function claims cannot be SE function claims. Neither can function claims be SE claims if we can point to the lack of alternative traits as an explanation for why better traits are not present. That kind of claim might be adduced if an investigator looks at our moths in GV in figure 5.1. If the moths really contribute to fitness only by providing camouflage, then there should be a better orange match present instead. But that can be defeated by suggesting that no other variants have yet arisen in the orange environment. If we can save the hypothesis that camouflage is a function of moth wings by denying that selection occurred, then function claims cannot be selection claims.

Another strength of NF is its fit with what biologists say about functions (Walsh 1996 makes a similar observation). Tinbergen (1963) famously distinguishes four questions for students of animal behavior, where the second question of function is given a nonhistorical slant. He tells us that when we ask, for example, why peacocks display their gaudy tails to mates we should distinguish four kinds of answer:

1. in terms of the physiological mechanism that produces the behavior;
2. in terms of the current functions, or survival value, of the behavior;

3. in terms of the evolutionary history of the behavior;
4. in terms of the development of the behavior in the life of the peacock

Krebs and Davies (1997, p. 4) endorse Tinbergen's view, telling us that we can answer the question of why a starling forages in a certain way: "In terms of function, namely how patch choice and prey choice contribute to the survival of the bird and its offspring." Godfrey-Smith (1994) reports a number of other biologists who concur with this view.

Finally it is important to see that for the purposes of biology it will often make very little practical difference which of the SE account, WT, or NF we choose. Most proponents of the SE account these days endorse some kind of "modern history" view of functions (e.g., Godfrey-Smith 1994; Griffiths 1993). That is, the function of a trait is the effect it was selected for in the very recent past. This move to recent history is made to preserve the distinction between functional and vestigial traits, where vestigial traits used to have functions but no longer do. NF has no problem with this distinction. Current functions derive from current fitness contributions, past functions from past fitness contributions.

What this all means is that in the actual world the functions picked out by NF, WT, and the modern history SE account are unlikely to diverge. For the vast majority of real-life cases, current contributions to fitness will also be recent contributions to fitness, which in turn will also be recently selected effects. Our imaginary cases in figures 5.1 and 5.2 yield different results for the SE account, WT and NF, only because we stipulate that no alternative variants enter into the population in figure 5.1. The population comprises only orange moths at the time the environment changes, and no other moth type enters the population subsequently. In reality this type of situation is extremely unlikely. Any real population will have some polymorphism at the time the environment changes, and even if it does not, some mutants will arise after that change. Just so long as some of these alternative traits are less fit than our orange moths in figure 5.1, then the camouflaging effect will be both an SE function and a WT function. So the SE and WT accounts will almost always agree in practice.

Moreover, if our modern history view is modern enough, then almost all NF functions will be SE functions too. A trait that makes a significant contribution to fitness will be an SE function just so long as some alternative, less fit trait with a similar effect was present with that trait in an earlier generation. Modern history accounts and nonhistorical accounts will only diverge in rare or downright implausible cases; the case in figure 5.1 where a novel contribution to fitness is provided by all the members of a population is highly implausible. Move to a time when there are variants in the population that are less fit, or move a generation forward in time, and the accounts will agree again.

Which function concept is the best for biology? Given the time it takes a biologist to investigate a population and make a function claim, all of modern history (SE) functions, WT functions, and NF functions will coincide. The dispute seems to boil down to one of whether the biologist should be most interested to make a claim about what a trait's current contribution to fitness is, or what its fitness contribution was a few generations back. It is hard to see how much could turn on this, yet I would cautiously advocate NF precisely because of its relative simplicity, because of its accordance with what biologists themselves say about the function concept, and because under NF a function claim can be established simply by showing a fitness contribution in a current population, without having to provide evidence to establish either that similar fitness contributions were made in earlier generations (as WT demands), or that alternative traits made inferior contributions in earlier generations (as SE demands).

5.8 Function and Design

I left one option dangling back in section 5.2. Isn't the idea of a functional trait one of a trait that has been shaped by selection for some purpose? Shouldn't we favor the etiological theory by recognizing (as Kitcher 1993 does) the link between function and design? And since design is a historical process—the design of an item must precede its use—then surely no item can have a function unless it has undergone the right kind of historical process?

We can follow Allen and Bekoff (1995) in sketching an informal notion of design in biology, where to be designed for some function is to have been subjected to a series of gradual modifications in the direction of improved function. This accords with an idea in the artifact realm that something can have a function in the sense of being used for some intended effect, without having been designed for that effect. A found rock can be used as a paperweight—hence weighing down papers is its function—but still it has not been designed to be a paperweight.

As suggested in chapter 4, the idea that selection "shapes" a trait—the idea that seems to motivate the notion that the design concept is appropriate here—relies on a form of gradualism in mutation. There will be no easy answer to the question of how gradual a series of adaptive steps needs to be to count as design. So the design concept will necessarily remain vague within biology, but I do not see that as a problem, since I find it unlikely that biologists would want to be able to give strict truth conditions for design statements in the way they might want for function statements.

This idea of the design of a trait corresponds well with another of Tinbergen's four questions—the question of the evolutionary history of a trait. Even when we establish the function of some trait—the contribution it makes to fitness—we say nothing about how the trait type came into existence. Now here is another role for the program of reverse-engineering. We can ask not only what the function of some trait is now, but whether it was designed for that function, or for something else. To ask this kind of question is to ask about selective history in a way that outstrips the selective history appealed to in the modern history view of functions. It corresponds to the difference between asking what problem a trait solves now, and what range of historical adaptive problems have shaped the form of an observed trait. Our moth wings in figures 5.1 and 5.2 have the function of providing camouflage, but this is not what they are designed for.

These considerations help to give some sense to Gould and Vrba's (1982) concept of *exaptation*. They contrast exaptations with adaptations: exaptations are traits that have been turned to some function, while adaptations are traits that are originally designed for some function. Critics have been keen to point out that on many definitions of adaptation, all adaptations are exaptations, and vice-versa. So if we understand an adaptation in Sober's sense as any trait that has been selected for some function, then no adaptation discharges any function that it was "originally" designed for, because evolution always proceeds by modification of ancestral structures. Even if an adaptation has been "shaped" for its function, no adaptation will have been shaped for that function originally—look far enough back and the trait will have been put to some other use. Hence all adaptations are exaptations. And even over a short period of time, so long as there is some variation in a population, it will be true to say that some novel function of a structure is also selected for. Hence all exaptations are adaptations too, except in freakish circumstances like figure 5.1 where there is no variation.

A workable distinction between exaptation and adaptation can be made, however, if we understand adaptations as traits that have been designed for their functions, while exaptations have not been so designed. The distinction will be fairly rough, for the standards by which we say a trait is designed for its function are also rough. However, there will be some traits that have undergone sustained cycles of gradual mutation and selection under constant selection pressures, while other traits simply enjoy success in a population that experiences altered selection pressures without any significant modification. We might choose to reserve the word "adaptation" for traits that fall on the "design" end of the continuum only. Note, significantly, that this distinction does not require that we subscribe to NF. The proposed separation of function claims and design claims can be made with SE and WT also. Even modern history accounts of function give camouflaging functions to the moth wings in figure 5.2 that the wings have not been designed for.

)

)eflating Function

.1 The Received View

Vhy is it that biologists use teleological language so often, but physicists nd chemists hardly use that language at all? It is true that these scientists ften use the word "function"—for example, a physical chemist may ask What is the function of free radicals in the breakdown of atmospheric zone?" Even so, physical scientists will rarely use "function" in concert vith terms like "purpose," "problem," "solution," or "design." So our riginal question remains. Why is it that the artifact model is applied in iology, but not in other sciences?

The selected effects (SE) account of functions, outlined in the last chap- er, is by far the most popular analysis of the biological function concept. 'his account also suggests an answer to the comparative question that tackle in this chapter. For the SE account, functions are grounded in election. So it is no surprise that only biology makes use of function talk, or only biology deals with systems subject to selection. This kind of po- ition, one that I suspect is as much a part of philosophical orthodoxy as he SE account itself, is expressed by Griffiths's (1993, p. 422) claim that . . . wherever there is selection, there is teleology." Griffiths also seems to nply through his selectionist account of artifact functions that wherever here is teleology, there is selection.

I will argue, based on the observations of chapter 2, that the artifact nodel only becomes practically applicable and psychologically attractive o inquirers when selection ranges over items with the right kind of de- elopmental organization. When items have this character, the result of

selection is the creation of systems with traits that have the kind o
functional complexity reminiscent of designed objects. In the contex
of systems like these, there is a psychological motivation to think of th
outcomes of selection processes as purposive—this is so because they ar
reminiscent of human design. There is also a strong attraction to thinkin
of selection pressures as "problems," hence to thinking of the whole evo
lutionary process by analogy to the creation of artifacts. In circumstance
where selection pressures do not act across these kinds of systems, th
heuristic payoff of thinking in terms of problems and solutions is so weak
and the resemblance to designed objects so slight, that the vocabulary o
function and design does not appear even though selection processes ar
in operation.

With this in mind, we should then ask whether there might be nonselec
tive processes that could ground artifact thinking in the physical science
provided, again, that they range across the right collections of objects.
argue that inorganic *sorting* processes could do this. I then confront a se
of arguments that might try to show that while these sorting processe
confer mere "as-if" teleology, selection processes confer genuine teleol
ogy. There is no relevant distinction that will tell us that only selectio
yields "true" teleological systems.

That is why I think of my account of artifact talk in biology as *defla
tionary*. One makes a mistake if, noticing the wide appearance of tele
ological language in biological inquiry, one assumes that there must b
some special process, which only organisms undergo, that bestows nor
mative, purposive states on them. There is a better explanation for the ap
pearance of teleological language in biological inquiry, which shows tha
any view that credits natural selection with a unique ability to bestow
norms is superfluous. Moreover, no account of why selection bestow
purposes serves to demarcate selection from sorting processes in thi
respect.

At the end of the chapter I show how the account given here resolves
number of apparent problems for a theory of functions. It shows that w
should tolerate multiple function ascriptions for biological items, and i
explains the continuity of teleological language in biology over the pas
250 years in spite of the change in the processes thought to produce such
systems.

2 Selection without Purpose

he place to begin our analysis of the link between selection processes
ıd teleology in biology is with the handful of counterexamples raised
r the SE analysis. I do not mean to say that these cases are fatal to the
E account; however, any theory of functions owes us an account of what
e should make of them.

I know of four examples where philosophers have tried to show that
though the conditions for selection are realized, investigators are not
clined to speak in terms of purpose. Outside of biology the example
f clay crystals has been invoked. Within biology the examples of the
nmune system, selfish DNA, and segregation distorter genes have been
iscussed. I outline each example in turn:

lay Crystals Mark Bedau (1991) offers an imaginary scenario based
n a hypothesis regarding the origins of life put forward by Graham
airns-Smith (1982). A dead planet harbors a population of crystals.
hese crystals reproduce, new crystals are formed through the seeding
ction of crystals already present. What is more, variations are heritable.
Iew crystals tend to keep the same regularities and irregularities of those
om which they are seeded. Finally, there is variation with respect to
tness. The crystals differ, and those differences make a difference to their
ropensity to make new crystals. Now suppose that some crystal has
defect D that means that it reproduces in sheets of a certain form, and a
am is formed against the solution supplying the raw materials for crystal
ɔrmation (just this example is discussed in Dawkins 1986). Here, crystals
f type D will tend to grow more rapidly because this damming action
nds to bring them more of the growth solution than non-D crystals.
think Bedau gets our linguistic intuitions right when he says: "...we
ould *not* want to say that, as a matter of natural teleology, the *purpose*
f D is to make dams, that D is there *for the sake of* making dams or *in
rder to* make dams." (Bedau 1991, p. 654)

he Immune System The dominant selectionist account of immunology
ells us that the immune system comes equipped with a vast array of anti-
odies. So there is ample variation. When some antigen arrives in the

body, the matching antibody is cloned in vast numbers. So there is repr duction with resemblance. And the match between antigen and antiboc determines which antibody will be most favorably reproduced. So there heritable variation in fitness among antibodies. Here I admit our inclin tions are not clear. Do we say that a specific antibody has the purpose matching the antigen? Do we say that the whole immune system has tl purpose of producing antibodies that match antigens? Do we say botl Here is how Mohan Matthen (1997) assesses the situation. He imagin clonal selection being undertaken in a test tube:

> Since proper function depends on selection alone, the isolation of antibodies fro any action taken by a system makes no difference—the proper function of tl cloned antibodies would be to match the triggering antigen just as before. B surely these proper functions do not transfer to the realm of teleological function With respect to what are we to reckon this antibody-antigen match functiona Outside of the system and its actions, the antibody is no more functional than tl Cairns-Smith crystals. (Matthen 1997, p. 29)

Selfish DNA/Segregation Distorters Some portions of DNA make copi of themselves within the genome, so that different copies of the sam apparently useless stretch of DNA appear at several different sites o chromosomes. Here there are properties in virtue of which some stretch of DNA are replicated more successfully than others, but often biologis decline to speak of these as purposive or functional traits, instead view ing selfish DNA as a paradigm example of a function*less* item. Similarl Manning (1997) takes up the example of so-called segregation distort (SD) genes, in order to argue against the SE account. These genes hav effects during meiosis that increase the chances that they (rather tha alternative alleles) will be present among the gametes which enter int fertilization. Manning says: "This seems to be a paradigmatic case c selection; having the trait of being a segregation distorter increases th chances of a bit of genetic material's being passed on through generation as compared with other genes without the trait... None the less, bio ogists do not typically regard SDs as having the *function* of disruptin meiosis" (1997, p. 75).

How can we explain why function talk is not used in these outline cases? The thesis I wish to defend in this section follows the broadl pragmatic approach of the last chapter: *Talk of functions, problems, an urposes appears in contexts where artifact thinking is both practical and sychologically attractive.* Artifact thinking is practical, or so I shall argue, here selection operates over systems with certain kinds of characteris-cs so as to produce objects reminiscent of human design, or where other rocesses—sorting processes—operate over large and diverse collections f objects. In such cases, these processes result in the production of ob-cts with many parts, each of which can contribute in a distinctive way owards some capacity of the whole. Such systems are thus reminiscent f objects produced by human designers, hence adding a strong element of sychological compulsion to artifact thinking.

The two sets of reasons (practical and psychological) are related. The evelopmental organization of biological systems, for example, explains hy selection pressures yield complex, modular, multifunctional systems hen they act on those systems. Hence the organization of biological ystems explains both why it is practical to think of selection pressures s problems to which fairly discrete solutions may evolve, and why the esulting systems themselves are reminiscent of designed artifacts.

Let us begin with the clay crystals. Why would we not think of clay rystals as having purposes? The answer, put quite simply, is that even ough clay crystals are subject to a selection process, they are not the right ind of item to evolve so as to acquire traits whose complexity gives the ppearance of design. We saw that teleological language in biology arises the context of projects like reverse-engineering, optimality modeling nd adaptive thinking. So let us consider how we would fare, were we to pply these techniques to a population of clay crystals.

If clay crystals undergo selection processes, then it must be possible at ast in principle to subject them to the techniques of reverse-engineering nd adaptive thinking; that, in turn, is to think of them in teleological rms. So we could posit a series of selection pressures for the crystals, nd ask which will be met, what constraints will affect the attainment of ifferent solutions, how different problems might interfere with each other produce complex trade-offs, and so forth. We can see how approaching e crystals in this way could tilt us towards thinking about them in terms f problems and purposes. We could also try to reverse-engineer a pop-lation of crystals, asking what functions different crystals parts have, hat problems they might have been designed to solve, and so forth.

The reason we do not approach crystals using the artifact model, is tha the artifact model would almost certainly be useless for the investigatio of crystals. Crystals are subject to a range of selection pressures—that i true. So we could ask why different crystal lineages do not engage in arm races with each other, why they do not produce progressively better an better dams, why they do not try to parasitize other crystals, or sabotag their access to nutrients. All of these are effects that we would expe might increase the fitness of crystal lineages, hence which we can thin of as selective problems that are brought to bear on crystal lineages. Y crystals do not evolve good adaptations in response to these selectio pressures.

We could explain the failure of crystals to develop complex adapta tions within the terms of the artifact model: crystals do, in fact, adopt series of extremely poor solutions to these problems, which are the re sult of highly complex trade-offs, ensuring that most parts of crystals ar compromise solutions that reflect the many functions to which they hav been turned. Yet the construction of such a response would be almos impossible. Trade-offs will be so wide ranging for crystals, and the rang of problems addressed so nebulous, that no successful enterprise of adap tive thinking or reverse-engineering could be undertaken. We will be abl neither to predict the types of crystals that will arise, nor to unpick th selection pressures that have acted on crystal populations. Interestingl it is precisely because Gould and Lewontin (1979) suspect that the orga nization of organic systems may be far more like the crystals than man biologists think, that they object to the attempt to save an optimal de sign analysis of the organism by recourse to the concept of trade-offs. will return to this neglected critique of artifact thinking at the end of thi chapter.

It is worth giving a brief reminder of why selection can fail to yield com plex adaptations. Suppose, for example, that the genome of an organism is such that the various loci controlling external pigmentation also contro many other traits. Suppose, that is, that were a genetic mutation to alte the color of a moth's wings, then such factors as the aerodynamic effi ciency of its wings or even the constitution of its eye would also change Then it will be clear that moths will not tend to develop camouflage, eve when there is a selection pressure for camouflage. Wing color cannot b

tered to meet the selection pressure in any significant way without that
recking the fitness of the whole organism. And if all the moth's traits are
und up with every other trait, then there will be no way to alter any one
them, without catastrophic effects on the moth's viability. The moth
ill not be able to evolve complex adaptations. So the basic point to be
ade is that only certain kinds of systems (systems whose exact charac-
rization I leave to others) are able to evolve by natural selection to yield
mplex adaptations. We can explain the fact that artifact thinking, and
nce teleological language, is not applied to clay crystals by appeal to
e fact that clay crystals have the wrong kind of organization to make
laptive thinking either useful or attractive.

Similar broad strategies explain our reticence to use artifact talk in the
her cases cited as counterexamples to the SE account, but here I will limit
y attention to selfish DNA and segregation distorter genes. We should
te, first, a pragmatic reason for not saying that the purpose of selfish
NA is to make copies of itself. It parallels the reason that we should
unhappy to say that the purpose of an organism is to make copies of
elf. Both claims are largely uninformative, since we could say the same
ing about all types of selfish DNA and almost all organisms. The claim
out the function of selfish DNA is not entirely uninformative, for it does
least assert the nontrivial fact of intragenomic selection. That is why
e find some biologists speaking of functions after all, yet not in tones
positive as we hear when they speak of the purposes and functions of
enotypic traits. One such quotation we have already seen: "If there are
ays in which mutation can increase the probability of survival within
lls without effect on organismal phenotype, then sequences whose only
nction' is self-preservation will inevitably arise and be maintained by
hat we call 'non-phenotypic selection'" (Doolittle and Sapienza 1980,
601).

As to segregation-distorter genes, Manning's assertion about what we
ould and would not recognize as functional is certainly up for debate.
e claim that these genes have the function of disrupting meiosis is
formative, and it begins to sound natural when we start to think of
gregation-distorters entering into a selection process with other parts
the genome for representation in future generations. Saying that the
nction here is to disrupt meiosis is not so uninformative as the claim

that they have the function of making copies of themselves—it says som
thing about how they achieve that. In that sense it is rather like a clai
about the biological function of a whole organism, one that gives som
kind of broad claim about how it makes its living. It is the genomic an
logue of a claim like "The function of cheetahs is to hunt large mammals.
What is more, it is not so strange to entertain the thought that biologis
would speak of the *parts* of segregation-distorter genes as having fun
tions. Just such a functional characterization is provided by Crow (197
in his discussion of the mechanism by which these genes disrupt meiosi

The general pragmatic story about why artifact talk enters into o
descriptions of natural phenomena works well across all of these exam
ples. Artifact thinking itself is useless for many selection processes, henc
artifact talk is not adopted. That said, when we start to think of items
undergoing selection processes in the context of environmental pressure
it becomes quite easy to imagine how we might start to attribute functior
to them. This thought is eased further when these items have the kind
internal organization that makes them apt for cumulative evolution.
these cases, the artifact approach is most fruitful, and the resulting iten
themselves grow to resemble the products of intelligent design more an
more. Hence we begin to feel more justified in our use of design concep
thanks to their heuristic value, and more comfortable in using them owin
to the resemblance between selected items and designed items.

Let me stress, finally, that this pragmatic approach to explaining th
peculiarity of artifact thinking to biology is really quite independent
the argument of the last chapter that NF gives the right sense of "function
within biology. One who is convinced that the SE or WT accounts wor
better still needs to face up to the apparent counterexamples discusse
here, and they should, I suggest, give the same response to those example

6.3 Purpose without Selection

Selection, we have just seen, is not sufficient for artifact talk, but is
necessary? What we should look for is a case where artifact thinkin
would be of use, yet where no selection process occurs. In this section
want to sketch some cases where what I shall call *sorting processes* migl
lead us to use teleological language.

A sorting process is one where there is variation across a collection of items, and differential propensities among the items to survive some kind of test, but no reproduction. The example of combinatorial chemistry with high-throughput screening, used in chapter 2, is helpful here. Drug companies discover useful drugs simply by putting millions of randomly generated molecules through a series of chemical "screens" where they are tested for some effect. The result is often the generation of a set of molecules well adapted to the tasks set by the screens, yet there is no selection process, at least not until the subsequent process of "lead optimization," in which successive modifications are made to promising candidates.

We might think of molecules that go through pharmaceutical screens as having functions or purposes even though they are not subject to a selection process. We might even take a set of molecules that survive the screens and subject them to a kind of reverse-engineering process. We could ask, given their nature and the effects they typically have, what kinds of screens they are most likely to have passed through. We can imagine thinking of these as problems faced by the molecules, hence we might think of molecules that solve those problems as having corresponding functions. So, for a collection of molecules, we might ask whether their purpose is to bind to some particular enzyme site, to break down some toxin, and so forth.

The case of the pharmaceutical screens is a real one, but it may not compel belief that nonselective processes can lead us to talk in terms of purposes and problems. That is so because pharmaceutical screens are designed by humans with certain purposes in mind. So one might say that the purposes we ascribe to these molecules derive not from the sorting process, but from the intentions of the scientists who designed the process.

That, I think, is a mistaken reaction to the example. True, these molecules have intended (IE) functions. Yet even had they not had such functions, we would still be able to use an artifact approach to infer the kinds of screens they had been through. So the example does suffice to show that, whether or not molecules have IE functions, the nature of the sorting processes these molecules have been through might be usefully approached from the perspective of the artifact model.

6.4 Will the Real Functions Please Stand Up?

In the last section I argued that we can imagine artifact talk arising in co
texts where no selection process is present, only a sorting process. Or
might think that the kinds of functions possessed by items that surviv
sorting processes are mock functions, or "as-if" functions at best. Co
sider a case like this: surface chemists are investigating how ions bon
to the surfaces of metal catalysts. If a collection of ions is supplied th
might bond to the metal surface, then a sorting process takes place. Thos
ions that maximize bonding stability will tend to ease out the ions th
are already bonded. In this case we can imagine how chemists migh
attribute purposes to parts of ions according to the role they play i
maintaining the ion on the surface of the catalyst. That would be esp
cially likely if our chemists discovered that the bonded ions themselv
have complex shapes, where specific parts can be assigned roles in ensu
ing adherence, and which prevent other competing ions from occupyin
those positions. Now although these chemists might *talk* about ions a
though they have purposes, one might complain that ions cannot *reall*
be purposive. Sorting processes are everywhere. Are we to say that th
intriguing "Brazil-nut effect" (Connor 2001), whereby large Brazil nu
work their ways to the tops of cereal packets, gives rise to nuts wit
ends? Surely we would not want to say that a process like longshore dri
bestows purposes on pebbles? Surely, that is, we would not want to spea
of pebbles in this case as having the purpose of being large in size, becaus
it is this that explains their survival through some screen?

Trying to discriminate between "real" and merely "as-if" functions
probably a waste of time. Here I want to argue that there is no nonarb
trary way for the proponent of the SE account to say why sorted functio
are any less genuine than biological functions. Again, that is why m
account is deflationary about artifact talk in biology—if selection ca
give genuine functions to eyes, then sorting processes can give genuin
functions to stones on the beach.

We cannot argue that sorting processes give rise to less genuine func
tions than selection processes, because both processes support the thre
connotations widely thought to be the marks of genuine teleology tha
I laid out at the beginning of the last chapter. Recall that teleologica

function statements are held to explain the presence of the functionally characterized item, to express normative demands on the item, and to allow a distinction between functions and "accidents." Not only can sorting processes yield functions that support these connotations, they support these connotations in more or less the same way as selection processes. We can see this by focusing on our ions bonded to the surface of the metal catalyst. The ions are, we imagine, in competition with other ions that might occupy their places on the metal surface. So the ions have effects that explain why they are there. Again, we can distinguish a number of senses in which they explain presence—they may have effects that, in the past, caused them to bond to the surface rather than some competing ion. This discharges the explanatory sense that is important to SE accounts of function. Or we might simply point to the effects that explain why they bonded, even when no competing ions were present. This corresponds to the explanatory sense that is important to WT. Or we might point to the effects the ion has that explain how it continues to adhere to the surface. This corresponds roughly with the explanatory sense of the NF account.

We might also give a criterion for when the ions are malfunctioning. We can do this by looking at cases where an ion, once bonded, is altered in some way so that it fails to adhere to the surface, or so that its adherence is weakened in some way. Here the ion fails to have an effect that helps explain its presence. That is just what it takes for a trait to malfunction on the SE and WT accounts of biological function. These theories say that a trait is malfunctioning when it fails to have the fitness enhancing effect that explains the presence of traits of that type. If there are ions of similar types bonded to the surface, and if one of them is altered to lose some effect that enhances bonding, then we can also say that the ion malfunctions in the approximate sense of NF.

Last, we can distinguish functions from accidents—at least in the sense that is achieved by the SE account and WT. An accident is an effect of a trait that enhances fitness without having explained the presence of the trait. Were a bonded ion to be disturbed in its configuration so that it begins to adhere more strongly to the surface, then this would also count as an accidental benefit in roughly the same way.

In summary, the kinds of function concepts that are supported by biological selection processes are no more "genuine" than the kinds of function concepts supported by inorganic sorting processes. If we feel that they are more genuine, then we can best explain that by pointing to the simple fact of historical habit (we are used to using teleological language in this domain), and to the fact that natural selection, operating over organisms, tends to give rise to entities that look more like designed items than do inorganic sorting processes operating over physical entities.

6.5 Two Kinds of Problem

Biological items have no stronger claim to the possession of genuine functions than do inorganic, sorted items. I have not said whether these claims are legitimate. Either stones on the beach and hearts both have genuine functions or neither does; but which is it to be? Most will think that because stones clearly do not have genuine purposes, then hearts do not either. Yet that might be a mistake if we think (like Griffiths 1993) that artifacts themselves—the paradigmatic instances of genuinely functional items—get their functions from selection or sorting processes also.

Here I want to show, quite simply, that the concept of artifact function that derives from intention is quite different from evolutionary concepts of artifact function that ground those functions in selection processes. Once this mapping of the different function concepts is achieved, no interest remains in saying which function concept is the "genuine" one.

We can see the difference between intended and evolutionary functions most clearly when we look to a case of an artifact that is also an organism—one that is subject to a natural selection process and which consequently has both types of function. So take a case of artificial selection, where a breeder has intentions to produce some kind of creature, and the creature itself is produced by an iterated cycle of selective breeding. In this case the creature is both an organism and an artifact, subject to function concepts grounded in both evolution and intention. Let me begin by distinguishing what we might call the design problems of the breeder—D-problems, for short—from the evolutionary problems faced by the creature—E-problems. An evolutionary problem, as we saw in chapter 4, is just a selection pressure. So suppose there is some selective regime such that cows that produce more milk tend to be allowed a longer breeding life, while those that produce little milk are sent to the abattoir early and have no calves. Here cows face the E-problem of producing more milk. Let's think of a D-problem as just what the breeder wants or intends from the breeding process. If he wants cows that produce high milk yields, then this is a D-problem also.

In many cases D-problems are reliably translated into E-problems. If a breeder wants to produce a high-yielding cow, then he should try to set up a corresponding E-problem for his cows. That helps to ensure that the right cows reproduce, and that a cow satisfying the original D-problem is produced at the end of a breeding program. In many cases, then, the production of milk is both an E-problem and a D-problem. These problems can come apart when the design process itself goes wrong—when the breeder fails to translate his D-problems into E-problems for the cows. A contrived example will make this clear. Suppose our breeder is unaware that the system he has for recording the milk yields of individual cows is subject to a series of systematic errors that ensure that the wrong yields are attributed to the wrong cows. High-yielding cows tend to be recorded as giving very little milk, and vice-versa. The decision on whether to send the cow to the abattoir is made on the basis of its recorded milk yield. Here we have a situation where there is no E-problem for the cows to give a high milk yield. The less milk they give, the more likely they are to survive and reproduce. Yet the D-problem of these cows—their intended effect—is still to produce milk, for this is what the breeder is trying to achieve with his system.

This example highlights a couple of points. First, even if we think that all artifacts are subject to selection processes of a sort, and that selection processes are what explain their form, that does not commit us to the view that their functions are grounded in those selection processes. In the case of our cows, a cow that has been subjected to a process of natural selection for low milk yield might have an intended (IE) function of giving large amounts of milk. If our aim is to predict the form of cows arising from the breeder's endeavors, then we are better off knowing the E-problems they face and not the D-problems of the breeder. Still, it is open to us to recognize a legitimate function concept that is grounded in D-problems and not E-problems.

Second, we can see that both organisms and artifacts will typically face far more E-problems than D-problems. Take a case where the selective regime for our cows is set up well, so that the high-yielding ones are recognized as such and the breeder gets what he wants—a high-yielding cow at the end of the process. Here there is, as we saw, a D-problem to maximize yield and also an E-problem to maximize yield. Yet the cows face myriad additional E-problems in their selective regime that are not D-problems. If cows can increase reproductive success by producing milk, then they can also increase reproductive success by contributing more than other cows to the milk tank, by causing the breeder to enter higher figures for yield into their charts, by causing the breeder to believe that they should not go to the abattoir, and so forth. All of these are E-problems faced by the cows, yet the breeder's own D-problems do not include that of making a cow that will cause him to believe it is producing lots of milk, nor that of making a cow that should cause him to enter high-yield figures into their charts. We have seen, then, that different types of problem, and the functions they yield, have different contours. Once we map these contours there is no interest in saying which kinds of functions are genuine, nor are there any resources left to answer this question.

6.6 Multiple Functions

The comments about the differences between E-problems and D-problems help to clear up a couple of remaining issues for our account of functions in biology. Take the case of the cows. Suppose we ask "What is the function of their large udders?" If we are talking about intended functions, we can give a univocal answer that is grounded in the intention of the breeder. We can say that their function is to give high milk yield, and we do not need to say also that their function is to make the breeder believe they are giving a high milk yield. Only the former is an effect that the breeder intends. If we ask instead about a biological function, then we must accept both claims, since both give fitness enhancing effects of the trait. Making the breeder believe the cows' milk yield is high is an effect of the udders that increases the cows' fitness. (For those not convinced that the NF account is right, the claim about multiple biological functions will arise under SE and weak theories also. See Goode and Griffiths 1995; Neander 1995c.)

If we have an urge to say that the real biological function of the cows' dders is to give milk, and that these others are bogus functions, then that only because we are implicitly running together the ideas of intended nd biological functions. Biological functions cannot pick out just one tness enhancing effect of a trait and claim privilege for it over other fitness nhancing effects that are either causally upstream or downstream from hat effect. What this gives us, I suggest, is an explanation for why in some ases where there is no designer, we are nonetheless inclined to think of ne of a trait's biological functions as privileged. We pick a privileged ffect because we continue to project an intentional designer into the ackground of the selection process, in such a way that seems to legitimate iving a single function ascription.

In most cases this kind of projection is quite harmless; indeed it may onceivably have practical benefits. The projection of a designer allows s to distinguish a manageable number of levels at which we can pick out tness contributions of a trait, by analogy with the subset of effects that xplains the proliferation of an artifact that are also intended effects of the rtifact. However, there is no basis in biology itself for arguing, say, that he function of the heart is to pump blood but not to bring nutrients to ells. The heart pumps blood, which in turn brings nutrients to the cells. oth are effects of the heart which contribute to the fitness of the organism hich bears it.

Not only is there no basis for unitary function ascriptions in biological rocesses, there is little basis in biological practice. It is important for bilogists themselves to distinguish different functions of traits at different vels of explanation. There are several examples in Raff's (1996) disussion of the functions of various genes, where the distinction between rimary functions and downstream functions is essential:

> he paradoxical story that is emerging from the discoveries of phylogenetically idespread regulatory gene families is that regulatory genes that are more or less imilar control quite distinct ontogenies. The wide range of possibilities in gene tilization allow this to occur. Conserved receptor-ligand systems can be linked quite different second messenger systems. Conserved transcription factors can roduce new patterns of gene action if transcription factor binding sites of target enes have been changed. (Raff 1996, pp. 359–361)

Raff's idea here is that the function of a gene can change if the genes ith which it interacts change, even when all of its primary products remain the same. This is a case where we should say that the immediate functions of a gene remain the same, yet the downstream functions alter. Those philosophers who have tried to argue that an account of biological function should deliver univocal functions for traits have mistakenly tried to make the analogy with intended functions too close; it is only the imagined projection of a designer that sometimes makes us suppose that a trait has a univocal function.

6.7 Purpose from Paley to Dawkins

The relationship between E-problems and D-problems also helps us to explain the continuity of teleological language from natural theology through to the present day. In many cases, I suggest, when we attribute unitary functions to artifacts we have in mind the D-problems that their designers confronted. Suppose we are faced with a group of cows, all of whom produce vast quantities of milk. We might view them and infer, quite naturally, that they had a breeder, whose D-problem was to produce a high-yielding cow. We infer from the makeup of cows to the effect intended by a designer; that is, we infer from the fact that producing milk is a biological function to the fact that it is an intended function. Suppose we learn subsequently that the cows in question were not the product of artificial selection, but of natural selection working without human guidance. They live in an area where pasture and water are in plentiful supply, and producing milk has not drained otherwise valuable energy resources. We will most likely still say that the function of the cows' udders is to provide milk, for the first step of our inference—that high milk production is fitness enhancing—still stands. When we learn that there was no background of intelligent-design, we still make the same function attribution as before, even though we attribute only a biological function instead of an IE function. That explains how it is that biologists these days make many of the same function attributions as natural theologians once did, in spite of the elimination of intelligence from the processes that are thought to lie behind natural design. There are, of course, a number of differences. Current biology sees a good deal more conflicting functions, through interspecific and intraspecific competition, than natural theology working under the principle that the creator is beneficent. We would also expect natural theologians to recognize many more traits as directed towards the benefit of humans, rather than towards the benefit of the organisms that bear them. On the assumption that the function claims made by natural theologians were claims about intended effects, we might expect current biologists to make many more function claims, since, as we have seen, intended effects will tend to be fewer than biological functions. This last difference is open to question, since for an omniscient, omnipotent creator we might assume that every effect that a trait has is an intended, foreseen effect, even though for human selective breeders only a subset of effects of successfully bred animals will also be intended.

We saw that E-functions and D-functions do not always coincide. Take the case where our breeder sets up the wrong kind of selective regime for her goals, so that although cows have the intended function of giving a high milk yield, their biological function is to give a low milk yield. Here we might be confronted by the cows themselves and a range of additional information about the breeder's stated goals, and we will be able to see that the cows have different intended and biological functions. If we know only about the cows themselves, then we may instead assume that their intended function was to minimize milk yield. In this case, the assumption that biological functions express intended functions when organisms are created by an intelligent designer leads us astray, because the selective regime fails to translate the intentions of the designer into selection pressures. Of course if the designer in question is an omnipotent God, we cannot suppose that he would ever fail to translate his intentions into a successfully functioning item. We will not assert, for example, that the intended function of a gazelle's stotting behavior is to scare predators, even though God did not, in fact, succeed in designing any gazelles able to scare predators. That fact again helps to explain the agreement between natural theology and modern adaptationism on many counts: on the assumption that creatures are designed by an intelligent, infallible designer, we will tend almost always to attribute intentions to that designer that are actualized in the designed item, hence which, in the case of biological items, tend to name biological functions of those items. Had the natural theologians not thought God infallible, then we should have expected much less agreement between the intended functions they named

for organic traits, and the biological functions currently named for tho
traits.

6.8 Functions and Design Revisited

In this final section I want to broaden the discussion again to look at som
further limitations and pitfalls of applying the artifact model to biolog
cal evolution. In the last chapter, I noted that a biologist may ask wh
some trait has been designed for—not merely, that is, what its functio
is, but how it has been shaped in response to environmental problems. I
truth, some sustained period in the evolution of a trait may be marke
by a variety of conflicting problems (E-problems, that is), partial solu
tions, adaptive pathways followed for a few generations to be resume
much later, and periods when the trait makes partial contributions t
many different problems. One can nevertheless look back retrospectivel
at the outcome of such a period of evolution and say that it is as thoug
there were some small number of underlying design problems being ad
dressed throughout the evolutionary process. Such a claim can only b
made through a metaphorical appeal to D-problems. The intentions of de
signers can remain constant even when the environmental demands place
on the objects they produce show very little constancy. A single sustaine
D-problem, for example, can coexist with whatever chaotic reality o
E-problems may face a lineage of artificially selected cows.

The idea of organic design can sometimes be quite misleading. Whe
designing an artifact, an artificer can construct its parts wholly indepen
dently of each other, and when design proceeds in this way there is n
illusion in the idea that particular design problems are brought to bear o
specific parts of the artifact.

Functional specialization, and the assignment of different design prob
lems, thus *precedes* the process of modification and testing in many case
of artifact design. In nature, functional specialization is instead a
outcome of the action of selection pressures across whole organisms
If organic parts address discrete problems, this is not a fact that precede
the process of modification and testing, rather, it is an effect of it. Still, th
illusion of design does provide us with a way of rationalizing organism
and dividing them up into specific functional traits, even when in reality

we should say that all traits are highly multifunctional. The anthropomorphizing projection of design and design problems onto the adaptive history of some trait, even when a distortion of the biological facts, can have the benefit of allowing the investigator to pick out particularly salient trends in the evolution of that trait. In other words, the perspective of intentional design can help biologists answer Lewontin's problem: "The decision as to which problem is solved by each trait of an organism is equally difficult. Every trait is involved in a variety of functions, and yet one would not want to say that the character is an adaptation for all of them" (Lewontin 1978, p. 164).

The approach outlined in this chapter suggests circumstances in which we might expect artifact thinking to break down even in biology. As we saw, selection must range over items of a certain kind if it is to result in the generation of complex adaptations. Unless traits are fairly functionally isolated from one another, for example, they will tend to interfere with each other's operation so much that no sustained increase in fitness in any one direction—no sustained "shaping" by natural selection—will be possible. Now although traits like the eye are themselves evidence that at least some aspects of some organisms' general structure is such as to permit complex adaptation, there is no guarantee that all aspects of all organisms' organization are like this. It may turn out that many traits are the result of so many conflicting trade-offs and constraints, and that they have contributed to the fitness of the organism in so many different ways over their evolutionary history, that any effort to ask what problems they might be designed to solve will be epistemically doomed. In other words, even if eyes are the result of a sustained response to a single set of selection pressures, many traits may be far more like our mutating, evolving clay crystals. Like the crystals, such traits may not be amenable to any kind of reverse-engineering project that seeks to cast them as designed responses to sustained environmental problems.

What kind of teleological language might we expect from a biologist studying such traits? I suggest that we would find an impoverished artifact vocabulary, where the investigator might be happy to speak in a piecemeal fashion of various current contributions of those traits to metabolic or other processes—happy, that is, to speak of functions—without speaking of the traits' design history, nor of the problems they have been shaped to solve. That may explain why some developmental biologists are so skeptical of the adaptationist approach. Thus Raff, again, points to the utter lack of functional specialization visible at various points during development:

> The phylotypic stage really represents a midpoint in development, at which the early embryo has blocked out the primary germ layers and the modules characteristic of late development are just beginning to appear. It is at this developmental midpoint that ultimately widely separated and distinct modules interact with one another in complex and pervasive ways. What could be more starkly illustrative of this than the observation of Jacobson that heart mesoderm helps to induce the vertebrate eye? (1996, p. 205)

At a developmental stage when so many different elements of the organism have such complex interactions, it would seem hard to take any kind of artifact approach to the history of its parts. An attempt to say what selection pressures have shaped such a system, or how that system would respond to selection pressures, will surely fail. As with any system that has little functional isolation, the conflicting interactions over time between parts will ensure that sustained adaptive responses will be extremely rare. Trying to predict what forms may emerge in the future, or trying to understand what problems have been faced in the past, will be futile. And even though we may argue that it is only at certain points in development that these questions become intractable for traits like hearts, there may be some traits for which no attempt to assign a design history can be successful at any stage of development. These are the good reasons why practitioners from certain developmentalist schools are skeptical of the value of artifact thinking.

7

Artifacts and Organisms

7.1 Technological Change as a Selection Process

I have spent the first six chapters of this book investigating similarities between the modes of explanation and description available to us when we approach organisms, and the modes of description and explanation available to us when we approach artifacts. I have shown how we can use functional language to describe both, how evolutionary problems can be conceived by analogy with design problems, and how there are benefits and risks in using techniques like reverse-engineering in both domains. What I have not done, but what I undertake in this final chapter, is an investigation of the claim that artifacts as well as organisms evolve by a process of selection.

Many students of technological change have found an evolutionary approach attractive (e.g., Basalla 1988; Mokyr 1990; Vincenti 1990; and contributors to Ziman 2000). Although my own focus in this chapter will also be on technology, the comments I make have some relevance to other stances on cultural evolution including evolutionary epistemology and "memetics" (e.g., Dawkins 1976; Dennett 1995; Blackmore 1999; Aunger 2000, 2002).

My strategy will be to distinguish the question of whether technology change is an evolutionary process, from the question of whether evolutionary models are likely to give us insights that nonevolutionary models could not have provided (see Amundson 1989 for a similar approach). I will suggest that at an abstract level technology change can be legitimately described as "evolutionary." However, I shall argue that this fact does not entail that evolutionary approaches to technology will revolutionize

the way we understand this domain. At present, there is no reason to expect the creation of robust, general models of technological evolutio
except at the most abstract, uninformative level. At the end of the chapte
I suggest some avenues of research that may, eventually, lead to th
discovery of richer evolutionary models for technology change.

7.2 Artifact Fitness

Let me note first that many who have been attracted to informal evolutionary models of technology change can make do with a very modes
model that says that technology change is a sorting process, not a full
blooded selection process. The importation to economics or marketin
of metaphors from ecology (e.g., Moore 1996), crude ideas of artifac
fitness, and so forth, can rest on the very simple idea that collections o
artifacts of different types are supplied to markets, and that some surviv
and others perish according to how well they "fit" that market. This is a
economic analogue of our example from surface chemistry in the previou
chapter. There we saw how if we supply a random collection of ions to
metal surface, after a period of time the composition of the ions remainin
on the surface will be well suited to the demands of that environment.

In these cases of sorting we can speak informally in terms of selectio
pressures that determine what kinds of items are likely to emerge ove
time in response to the makeup of the environment. We can also speak
informally of the fitness of different products, in the sense of their suit
ability to the local ecological demands. We can talk of the different niche
of different artifacts, and we can make sense of the idea that niches are
constructed. That is because both in the case of the ions and in the case o
the artifacts, the nature of the environment that determines what kin
of entities fit and what do not will turn in part on how those very ions o
artifacts affect this environment.

Much of the mathematics of evolutionary biology, as well as its language, can be used in cases of mere sorting, whether that is sorting of
ions or of artifacts. For some particular type of ion we can assign a valu
that expresses the expectation that it will increase its concentration on a
metal surface over time. Such a value might be thought of as a kind of
trait fitness, or selection coefficient. And in the case of small populations

of ions (or of artifacts) we will also expect to see the kinds of small sample fluctuations that characterize genetic drift in the biological realm. That is to say, we will not expect these trait fitnesses always to translate perfectly into the actual changes in frequency of some artifact type, or some type of ion. What all this shows is that we cannot take either the appearance of the vocabulary of evolutionary biology in describing some nonbiological system, nor the usefulness of the statistical mathematics of population genetics for predicting the behavior of such a system, always to be indicative of deep similarities between the system in question and evolving natural populations.

A meatier form of technological reproduction does exist in the context of crafts, or primitive tool use. We can imagine a society in which there are two or three variant types of stone tool, and one type of tool is better at skinning animals than others. The good skinners are seen by observers to do that job well, and similar looking tools are made, while the inferior types are thrown away. In this kind of case, it is quite straightforward to think of a stone tool as an evolving entity. A tool type is copied by observers according to certain criteria, which may be functional, ornamental, and so forth. These criteria will partly determine the chances that a tool will promote the production of further tokens; that is, the criteria determine the tool's reproductive fitness. Some tools will be copied more often, while other tools which do their jobs poorly will be discarded. So long as the copying process is fairly faithful we can see that some tool types will tend to increase their frequencies in the population of tools according to these reproductive fitnesses.

The example we have just considered is one where the fitnesses accorded to artifacts reflect the propensities of these artifacts to promote the production of resembling tools, hence it is a situation in which these artifacts can be said to have offspring of sorts. This is not to say that the makeup of the "parent" tool is what wholly accounts for the makeup of the tool that is copied from it. Perhaps what happens is that the artisans are familiar with a small number of culturally endorsed methods for tool production, and the perception that one tools does the best job triggers the execution of a method known to produce tools of that same type. In this case, the resemblance between the original tool and the tool made in its image is accounted for primarily by cultural resources external to the artifacts themselves. Even so, the attribution to the tools of the property of reproductive fitness is clearly better grounded here than in the case of the metal ions.

Although an evolutionary model may seem defensible for these hand-made artifacts, we might worry that it cannot work for mass-produced artifacts. Take the marketplace for computer operating systems. There is certainly variation here—Microsoft Windows exists in a number of forms and (although this may come as a surprise to some) there are other variant operating systems too, such as Linux. These artifacts can be thought of as having fitnesses, where fitness is understood as a mathematical expectation to appear in future artifact generations. Just as organismic types vary in frequency across generations, so do operating system frequencies. It is because creatures of one type are better suited to their environments than creatures of some other type, that the former outcompete the latter to increase their frequency. Similarly, it is because MS Windows is better suited to its environment than Linux that the former outcompetes the latter and increases its frequency in the marketplace. Note that this does not commit me to the thesis that MS Windows is superior to Linux in any everyday sense of the word. Perhaps MS Windows is fitter in part because the environment for operating systems is composed of many computers already running MS Windows instead of Linux. Perhaps, that is, Linux is in the more usual senses of the word "better," but frequency-dependent selection results in MS Windows outcompeting it. Or perhaps what makes MS Windows fitter is tied to some perception of the MS brand—a perception that may not reflect the true qualities of the product. To say that one product is fitter than another in some marketplace does not entail that the first product is superior in an engineer's or a connoisseur's sense of the word—only in a marketer's sense.

The problem of applying the evolutionary model to mass-produced artifacts becomes apparent when we notice that successive "generations" of artifacts typically do not give rise to each other through chains of reproduction, but instead owe their production to a common cause. A batch of Minis in 2003 is not produced by cars from 2002: rather, both batches of cars come from the same production line. These token cars can be assigned reproductive fitnesses with greater legitimacy than token ions in the surface chemistry case. The token ions do not determine how many future token ions will be produced. On the other hand, the properties of token cars in one year do in part determine the number of cars produced in the following year. Even so, the causal relations between tokens of the different generations are extremely complex. Users of cars who are pleased with the product in 2002 might tell their friends about them, they will show them off, and so forth. This may lead to a demand that, mediated by the manufacturer's market research department, will result in increased production in the following year. So token cars in one generation have a causal influence over the number of cars produced in the next generation. There is, however, no possibility of tracing lines of descent between individual cars of distinct generations.

The contrast between token artifacts and token organisms is now quite clear. In the case of a token sheep, say, it is easy to say just which are the offspring of the sheep in question and what the chances might be of some sheep producing a certain number of offspring. Since assigning offspring to token artifacts will often be impossible, so will estimating these kinds of reproductive fitnesses for individual artifacts. It is not always so hard to say which are the offspring of which artifacts; cases where artisans copy their work from each other will be far easier to untangle. Even so, once we have noted this limitation for extending an evolutionary model to mass-produced artifacts we do not need reject the evolutionary analogy, so long as we are clear about what the fitness concept amounts to in the technological realm. Even in organic populations analogous problems arise for the interpretation of some fitness concepts. *Inclusive fitness,* for example, does not reflect the propensity of some token organism to produce offspring of its own. Rather, the inclusive fitness of some organism measures the number of copies of its genes for which it is responsible. Hence inclusive fitness includes the contributions an organism makes to bringing organisms into existence that are not its own offspring. We have seen that token artifacts are causally responsible for bringing copies of themselves into existence in future generations. So artifacts can be assigned inclusive fitnesses, even if they can rarely be thought of as having reproductive fitnesses.

Richard Lewontin has claimed in conversation that models of cultural evolution have difficulty coping with the phenomenon of power. My discussion of artifact fitness helps us to see why that should be so. So long as artifacts fluctuate in their frequencies in reliable ways, then we can assign numerical values to different artifact types that reflect their expected rates of increase over time. Artifacts of all types do this, yet we should be wary of thinking of these values as reproductive fitnesses even though they can play a similar role to organic reproductive fitness in predicting frequency changes of artifact traits over time. In some cases, the rate of change of some artifact type does not owe itself to the superior reproductive output of that type in virtue of its suitability to a market environment. Rather, if the artifact's manufacturer is sufficiently powerful, that manufacturer may simply elect to flood the market with the item in question. In such a case, we could use a mathematics that assigns expected rates of increase to different types of artifact to calculate how some type of car, say, should change its frequency over time. Yet this kind of case has very little in common with a population of organisms of different types, in which the performance of token organisms with respect to their environments helps determine the rates of production of further tokens of those types.

7.3 Where Are the Replicators?

The phenomena of power are genuinely problematic for an evolutionary theory of technological change. Setting these problems aside, so long as we do not insist that an item can possess the property of fitness only if it has some number of identifiable offspring, then artifact populations can be said to possess heritable variation in fitness, and they can be thought of as undergoing selection processes as a result. The next question to ask is just what role artifacts have in this selection process. Thinkers on evolutionary approaches to technology have been preoccupied with the question of what we should focus on as the units of selection when we think of technology change. Should the basic unit of analysis be the artifact, the idea, the process, the artifact/idea complex, the meme, or some other item? I will discuss this first using Hull's distinction between replicators and interactors. Although in the end I shall argue that the distinction is ill suited to the cultural domain (and some, e.g., Griffiths and Gray 1994 and Gray 1992, have argued for similar reasons that it is ill suited even to biology) the distinction is a useful starting point that helps us untangle the comments of others. Hull's definitions are as follows:

> *Replicator*—an entity that passes on its structure largely intact in successive replications. *Interactor*—an entity that interacts as a cohesive whole with its environment in such a way that this interaction *causes* replication to be differential. (Hull 1988, p. 408)

Typically, genes are thought of as replicators while organisms, and sometimes groups, are thought of as interactors. The search within the technological domain for analogues of genes and organic traits is an expression of the thought that distinct replicators and interactors should be found here also. The replicator/interactor distinction is of limited value, I shall argue, because, depending on changeable contextual variables, all of artifacts, ideas, artifact/idea complexes, and so forth, can act as replicators.

We need to begin by looking in more detail at the replicator concept. Hull's use of the active voice in the claim that a replicator "passes its structure on largely intact" is intended to exclude many items whose structures are merely *passed* on intact across successive generations. The distinction can be made clearer by means of an example. Cows produce cow pats that resemble the cow pats of their parental generation. So cow pat structure is inherited across generations. But cow pats are not replicators. Cow pats in one generation resemble those in the parental generation because the cows which produce them resemble the cows that produced parental pats. The structure of the parental cow pats themselves is not causally involved in generating the similar structure of offspring cow pats. Yet this is not to deny that cow pats could function as interactors. If some cows with especially unpleasant smelling dung are sent to the abattoir before they have the chance to have calves, then it is because of the nature of the cow pats that certain replicators (genes, say) are passed on to the offspring generation.

I suspect some hard line "developmental systems theorists" will disagree with my categorization of genes and cow pats as replicators and interactors respectively. No matter. The point is to see what function the replicator/interactor distinction is supposed to have, even if it turns out to have no application. The developmental systems theorist might argue that many portions of DNA do not fit the replicator criterion after all; perhaps cellular "proof reading" machinery will ensure that alterations are weeded out, and not copied. She might even try to argue that cow pats affect the pasture on which the cows live, which in turn affects the cows' diet, which in turn affects what kind of cow pats are produced. So perhaps cow pats do have a limited role in their replication.

We can now see why a reasonable test to see whether something is a replicator is to ask whether, were its structure to be changed in some way, that altered structure would appear in the next generation. Take out a portion of DNA in a germ line cell, and replace it with another, and that same type of DNA will appear fairly reliably in future generations. Take a cow pat from behind a cow and replace it with another, and that will have no effect on what kind of cow pats future generations of cows produce. The test provides evidence in favor of the view that it is the structure of the element in question that is, at least in part, responsible for ensuring the resemblance in structure across generations, hence that the item is, indeed, a replicator.

Note four things about using this test to determine what entities are replicators. First, the discussion tells us that we should call an entity a replicator only across a certain range of contexts. If we remove a portion of DNA from a germ line cell and replace it with another portion of DNA, then that change might be reproduced; if we replace it with a completely different type of molecule, or a live elephant, or perhaps some stretches of alien DNA, then daughter cells will not show that change.

Second, it is fairly likely that, among cellular and extracellular machinery, more than just stretches of DNA will turn out to be replicators. If parts of cell walls in paramecium are altered, then cell walls with similar structures will tend to appear in daughter generations (Nanney 1968). Even in sexually reproducing organisms, including mammals, alterations to structures such as methylation patterns can be reliably reproduced across generations (Roemer et al. 1997; for a full survey of these so-called "epigenetic inheritance systems" or EISs see Jablonka and Lamb 1995).

Third, before we conclude that an item is a replicator because it obeys the counterfactual test, we need to ask exactly how the item is to be altered. If noses are altered by altering the zygotic DNA that gives rise to them, then these alterations will be passed on in future generations. Yet noses are not replicators. The important question is whether resemblance in nose structure from generation to generation is caused by the structure of noses. Altering their structure is intended to give evidence about this, but it is not foolproof.

A final point, reinforcing the third, is that obeying the counterfactual test is compatible with the thought that having some structure *triggers* the production of a new item with a similar structure, yet does not serve as a *template* for it. Here is an analogy from Sperber (2000) that shows the difference. Imagine two scenarios, in both of which ten tape recorders are arranged in a line. In both cases, the first recorder in the line plays a randomly chosen tune. Now in the first situation, the next recorder in the line records the tune just played, and plays it back. The next recorder records that playing, and plays the tune again, and so on. In the second situation, the first few notes trigger a search through the memory of the next recorder in the line for a matching tune, which is played back. This playing in turn triggers a search through the memory of the next recorder, which also plays the tune, and so on. In both cases the tune is reliably reproduced by each of the recorders in the line. And in both cases if one were to change the tune played by the first recorder, the tune played by the next recorders would change to match it. So the second scenario is a case in which the tune obeys the counterfactual test, yet the structure of the tune plays a minimal role in ensuring resemblance from recorder to recorder, and certainly a far smaller role in this function than does the structure of the tune in the first scenario. Is the tune a replicator in this second scenario? That is a question I prefer not to answer. It is enough for our purposes here that we note first, in neither scenario is the tune "self-replicating" in the sense of being wholly causally responsible for its faithful reproduction. In both cases the same tune is played several times only because of the action of a great deal of additional machinery. And we should also note that the means by which faithful reproduction occurs are different in each case. The structure of the tune plays some causal role in ensuring that further tokens also have that structure, even in scenario two. But the tune does not act as a template in this case.

These results have relevance to the question of how we should understand technological evolution. First, take an example where there are no artifacts involved. Suppose we are instead considering some mental state M that produces a behavior B. Say the mental state is the desire to smoke a cigarette, and the behavior is smoking. Now it may be that under certain circumstances, say in a restaurant, if one individual lights up a cigarette others will follow. And they may follow, producing B', because they also acquire the desire to smoke M' as a result of perceiving the smoking behavior B. So we have a causal chain going from M to B, to M' to B', to M'' to B'', and so forth. What is the replicator here and what is the interactor? Is the mental state M the replicator? It seems so, since it passes the counterfactual test: had the mental state been different, then it may have resulted in that different mental state being copied. Perhaps had our smoker desired to smoke a cigar, or a pipe, then this mental state would have been transmitted instead. Note that for some substitutions, no copying would have occurred—had our agent desired to take his trousers off, this desire may have remained with him—but this kind of sensitivity of replication to what is substituted is equally observed for the replication of genes.

It would be easy to think of mental states as analogous to genes because mental states, like genes, can be thought of as "internal," "controlling" states with behaviors or artifacts as external "vehicles" by which they ensure their replication. However, we should not be seduced by the fact that mental states are "inside" agents into thinking that mental states are the *only* items in cycles of cultural reproduction that could merit the name "replicator." Mental states are not the only elements of these reproductive cycles that can be thought of by analogy with genes. In the example we just looked at, the smoker's behavior can also be thought of as a replicator in its own right. If the behavior were different in a certain way, then it may have been reproduced in that same way in other smokers. Had the smoker lit the cigarette with a certain flourish, or had he blown smoke rings, then that may well have appeared further down the room. Smoking behavior obeys the counterfactual test, too. Debates about cultural evolution which suppose, even in a specified context, that there will be one single type of item that is the cultural replicator make a false presupposition. There are a great many cultural replicators, understood as items that obey the counterfactual test.

The point is shown quite economically in figure 7.1. The top half of the diagram shows the traditional conception of molecular Weismannism; only gene/gene relations explain heredity, and only genes are replicators. Phenotypes may affect how many replicators of some type are produced, but they are not replicators in their own right. The bottom half of the

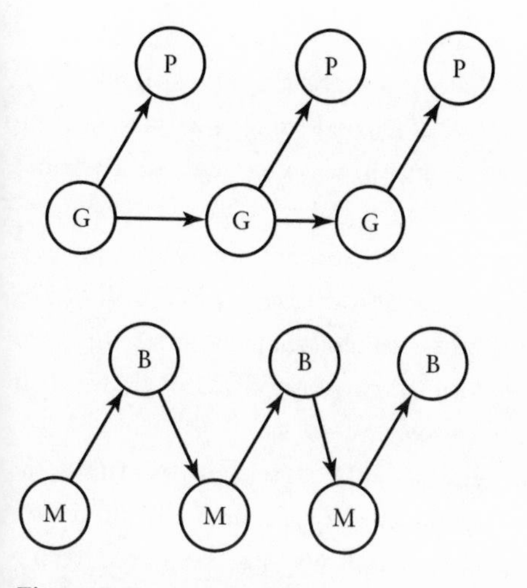

Figure 7.1
Instances of cultural reproduction (bottom), where mental states (M) are replicated through their effects on behavior (B). This cannot be represented by analogy with traditional representations of Weismannism (top), where causal relations between genes (G) explain heredity, while phenotypes (P) affect reproductive success.

diagram shows the relations between mental states and behavioral states in our imaginary example. Some critics of Weismannism argue that the top half in fact fails to represent biological processes accurately. Whatever the outcome of that debate, it is clear that the relationship between mental states and behavioral states cannot be understood by analogy with traditional Weismannism. Still, even if cultural reproduction is likely to be mediated by a variety of different types of replicator, we still have not said whether artifacts can play this role in technological reproduction. Are artifacts replicators?

My answer here is a cautious "sometimes," although again with a wary eye on qualifications about the possibility of very indirect copying, about the failure of the counterfactual test to distinguish template copying and triggering, and about the relativization of replication to certain kinds of substitution in certain contexts of copying machinery. We can consider a number of examples where alterations to artifacts are reliably preserved across generations. In any case where an artifact acts as some form of template for the production of other artifacts of the same type, then alterations to artifacts will tend to feature in future generations. Processes

of this type may occur in some artisanal contexts. Also, contexts whe
an item is observed and then reproduced, perhaps through some know
edge of the processes that went into its construction, allow us to see th
artifact as a replicator, albeit one whose replication is mediated by som
extremely complex processes.

In many cases, however, artifacts function as interactors without func
tioning as replicators. We saw that properties of tokens of MS Window
affect the number of future MS Windows tokens that will come into ex
istence; however if I make an alteration to my copy of MS Windows,
will tend not to pass its structure on to other copies.

Remember also that whether an alteration is passed on to future gen
erations depends, for organisms or artifacts alike, on how the alteratio
is brought about. If my version of MS Windows is altered by alterin
Microsoft's production line, then these alterations will also be passed on
It is possible, all the same, that different ways of alterating MS Window
tokens could pass the counterfactual test for being a replicator under cer
tain circumstances. This will be the case, for example, if I pirate copies o
my modified software and send it out to other users. Or I could commu
nicate my innovation back to Microsoft and Microsoft might decide t
try my new variant in a production run. And if Microsoft have a standin
policy of monitoring all alterations to extant copies of their software, an
releasing test versions of those variants, then again these artifacts pas
the counterfactual test for replication—alter them and those alteration
will appear reliably in subsequent generations. Yet in general, for mos
alterations, most copies of MS Windows are not replicators.

The answer, then, to the question of whether artifacts are replicators i
that artifacts of all types can sometimes be replicators in some contexts
Sometimes, however, they act as interactors without also acting as repli
cators. And in many cases even when they act as replicators they are als
involved in the replication of other items such as beliefs or manufacturin
processes.

In summary, the question so pressing to many students of cultura
evolution—What are the proper units of cultural selection?—is likely t
be a bad question to ask since:

1. It glosses over the replicator/interactor distinction. Even if artifacts ar
interactors, it does not follow from this that they are not worthy of clos

attention, just as noses may be worthy of evolutionary scrutiny even if they are not replicators.

2. Whether an artifact acts as a replicator will depend on contextual factors. There is unlikely to be any short answer to the question of whether all artifacts are replicators or not. And the stability or fecundity of a single type of cultural structure across long periods of time could be due to the replication of different types of element at different times. Hence trying to focus on any one type of entity may have the effect of distracting attention away from how artifacts and practices are preserved, disseminated, and modified by imposing a false monism on the items responsible for cultural inheritance.

This second point is, I think, the most damaging to the ambitions of models of technological evolution. Causal chains of technological reproduction are typically mediated by agents with the capacity to observe and reason abductively. Elements of the causal chain of reproduction are replicators just when they obey the counterfactual test: were they to be altered, then those alterations would be passed on. It is in virtue of the attention of artificers, market researchers, consumers, and regulators that otherwise inert elements of production can come to satisfy that condition. If an observer has her eye on a part of a car, then that part can become a replicator. Absent this attention, it does not pass the test.

What this means is that for any given area of technological reproduction, the shifting attention of agents involved in the reproductive process will cause different items to become replicators at different moments. Worse, chance alterations to different types of item may cause agents to pay attention to them, thus resulting in their becoming replicators for a short time. There is little chance of finding any stable series of replicators, hence little chance of establishing a general, informative, theory of cultural inheritance, in virtue of the role of reasoning agents in the processes of technology change.

7.4 Is Technological Evolution "Lamarckian"?

One often hears the claim that no theory of evolution by natural selection can be applied to the cultural realm because cultural evolution, be it conceptual or technological, differs from organic evolution in that the former is Lamarckian while the latter is not. But the vocabulary of Lamarckianism tends to hinder the debate more than it helps it.

What does it mean to say that an evolutionary process is Lamarckian? I do not want to look to history to unearth what Lamarck intended by his theory; however, it is clear that in the modern debate the term has acquired two very different meanings which are often run together. The first regards the question of whether changes acquired during the development of one generation are inherited in daughter generations. The second regards the possible directedness of variation. Both of these senses stand in need of further disambiguation but for now it suffices to note how different they are from each other. It is possible, for example, that genetic mutations tend almost always to produce beneficial mutations. In this sense we might say that genetic mutations are not blind but directed. This says nothing about whether evolution also proceeds by the inheritance of acquired variation. It could be that giraffes living in areas where foliage is just out of reach of their necks tend to give birth to offspring with mutations directed in favor of longer neck length. Still those giraffes which strive during their lifetime to reach the higher trees, and whose necks grow longer as a result, may not pass these changes on to their offspring. So there is no inheritance of acquired traits. Equally, if acquired variation is passed on to offspring, this is not to say these changes are directed. The two senses of Lamarckianism are entirely independent of each other.

The notion that acquired variations are inherited is itself rather hard to pin down. A strong sense of this kind of inheritance seems to have emerged in recent years (not one that Lamarck could possibly have intended), that equates the inheritance of acquired variation with a process whereby beneficial modifications to the phenotype during the course of an organism's life could somehow be read back into its germ line DNA. So far as I know, there are no cases thought to exist of Lamarckian inheritance in this strong sense.

Wary of the pitfalls of the term "Lamarckian," what should we say about technological evolution? First, are inherited variations acquired? This, I think, is a matter of perspective. It is hard to see how the strong sense of Lamarckian inheritance could apply to technological evolution. What would it mean to say that phenotypic changes are read back into the germ line in the case of technological change? In the last section we saw some serious difficulties in saying what it might mean to speak of the germ line in this domain.

At best, it would seem to mean that alterations to an artifact, say, would be read back into the processes or ideas that initially gave rise to it. So we might imagine drawing a blueprint for an artifact, building the artifact, finding some accidental and useful change occurs to the artifact during its lifetime, and then revising the blueprint of new artifacts to reflect this change. This kind of change occurs quite often; however, our assessment as Lamarckian in the strong sense will tend to depend on what we have chosen to focus on as replicator. Consider the case once more of epigenetic inheritance systems like methylation patterns. If a germ line methylation pattern is altered, then this change can be preserved in later generations. This is to say that the processes which go into the production of the methylation pattern must be altered also. So we might say, in a very weak sense, that these changes are Lamarckian. However, we could just as easily say that changes to germ line DNA are Lamarckian. If a nucleotide sequence is altered in some way, this will be preserved in the next generation. If we see the DNA sequence as the "phenotype" for DNA reading processes, then this begins to look like Lamarckian inheritance. An alteration to the DNA has been preserved, and this has been achieved by some alteration of the processes that gave rise to the DNA. If we are already disposed to see DNA stretches as replicators, then we will not see the recreation of items with similar structure in subsequent generations after alteration as Lamarckian; otherwise, we will. Equally, if we are already disposed to see artifacts themselves as replicators, then we will tend not to see the recreation of items with similar structure in subsequent artifact generations as Lamarckian. Since, as I have suggested, replicators are likely to be ubiquitous at many levels in artifact evolution, the classification of technological evolution as Lamarckian tends to stumble in similar ways to the replicator/interactor distinction on which it relies.

In spite of all this, there is something importantly right that lies at the heart of the view that technological and biological evolution differ in that only the former is Lamarckian. True, both organisms and artifacts are reliably recreated through the action of complex, interacting inheritance systems. Yet as the argument of the last section showed, our capacities for reasoning mean that technological reproduction is not merely a domain

containing multiple inheritance systems, but one where any alteration, t
any type of element of a system, can be coopted to the "germ line." Bette
perhaps, to use the shifting germ line itself as illustrative of the problem
in using an evolutionary approach to culture, than to argue that acquire
mutations can be inherited in the germ line.

7.5 Evolutionary Explanations

Although only loosely related to the notion that technological evolutio
is Lamarckian, we need some investigation of the assertion that unlik
biological evolution, technological evolution is directed. Again, the con
cept of the directedness of evolution is not without ambiguity, althoug
this time I am more sympathetic to those who assert a strong disanalogy
I will look at the disanalogy first, then move on to consider how thi
might compromise the explanatory virtues of an evolutionary approac
to technological innovation.

What does it mean to contrast mutation that is blind with mutatio
that is directed? The distinction is tricky to make precise, and it is no
a task that I need accomplish here. No mutation is wholly "random"—
indeed the meaning of randomness is also hard to make precise. Eve
genetic mutation is not random in the sense that all items are equally likel
to arise through mutation at any locus. As I already mentioned, singl
base substitutions are precisely that—substitutions of one nucleotide fo
another—and in this sense what kind of entity appears through mutatio
is not a random matter. What is more, some types of mutation may b
more structurally stable than others. Genetic mutation can be directed i
this sense without posing any threat to the traditional Darwinian view o
mutation, since the fact that some mutations are more likely than other
says nothing about whether it is the fitter mutations that are more likely

I will take the claim that mutation is directed to mean that there is
bias in mutation in favor of mutations that increase fitness rather tha
decrease it. To say that technological evolution is directed is to say tha
changes made to an artifact will tend to be "sensible," whereas change
made to a genetic system will not. So, when we imagine a composer trying
to perfect a piece of music, it seems unlikely (unless she is a surrealist) tha
she proceeds by substituting random notes or phrases and waiting unti

omething turns up that sounds right. Instead, it seems, the composer will nore likely choose from a subset of notes or phrases that she believes vill tend to make the piece better. In this sense it seems that variation is ndeed directed in artifact evolution.

How far does this undermine the Darwinian view of technological innoation and technology change? I think we should make three preliminary omments. First, we need to remember our two senses of "directed." If turns out that our composer does not substitute just any note in a iece, but instead looks to a fairly tightly circumscribed set of options, nis shows that variation is directed in the first, weaker, sense of there eing some bias in which variants are likely to arise; however that does ot show that the variants are all more likely to improve the piece than nake it worse. Second, if variation is directed in this second, stronger ense (as I grant it may well be), then this does not destroy the analogy vith organic evolution entirely. As Jablonka has tried to show, it seems hat many mutations in organic evolution may be directed in precisely the ame sense. Third, and most important, even if variation is directed, this oes not deny an important explanatory role for Darwinian selection. A elective mechanism can still explain why, of a set of variants, the fittest mong those variants was chosen, even if the set as a whole is directed. In ther words, there may still be significant fitness differences between the nembers of a directed set. Selection processes may play a variety of roles n explaining innovation even when innovation is directed.

.6 An Evolutionary Revolution?

hope it is clear from the argument so far that evolutionary approaches o technology, while useful organizing frameworks for investigators who ike to think in the language of biology, have not yet given us much to ope for in terms of a revolution in our understanding of technological hange. That is the case even if the assertion that technological change is n evolutionary process captures an important truth.

But what does the evolutionary approach have when applied to organisms, that it lacks in the technological realm? Population genetics has n advantage over any population approach to the study of technologial change in the fairly well understood patterns of inheritance for many

Mendelian traits. An understanding of the genetic basis of a trait can counter predictions of frequency change for that trait based on estimated fitness if, for example, we know that the fittest trait is coded by a heterozygote (Sober 1992). Here our understanding of the underlying system of inheritance allows us to predict that the fittest trait will not go to fixation, since less fit traits will always be reproduced by segregation. In other words, what we might naturally predict on the basis of the fitness of some trait in an environment can be overturned by adding additional assumptions regarding how the trait is inherited.

Where evolutionary models for technology change fall down is in the lack of any analog to transmission genetics. I have suggested that we will probably find no context-independent answer to the question of what the replicators are in technological evolution. We can add to this the fact that there is no uniform system of reproduction in technological evolution. Sexual reproduction involves the contribution of two parents to form a new organism. A new token artifact can be the product of the union of as many, or as few, models as the artificer happens to draw inspiration from. An artisan might make a tool by combining elements from any number of different stone tools, or from a stone tool and a wooden tool. One often hears the slogan "Memetics needs a Mendel," yet the ways in which technological inheritance is ensured, and the many ways in which technologies combine with each other, are likely to depend on fine-grained contextual factors in a way that will make the discovery of any general rules of technological transmission very unlikely. Of course the biological world, too, does not contain just one kind of reproductive system, and Mendel's laws are frequently broken. Yet it seems unlikely that technological change will show the kinds of islands of homogeneity that we would need even for useful rules of thumb in transmission to take hold.

Enthusiasts of evolutionary models should also note that many traditional explanations of technology change are likely to remain largely untouched even if evolutionary thinking becomes popular. To say that a certain technology succeeds owing to great fitness, or to rank artifacts in terms of their fitnesses gleaned from studying rates of incidence in a population, does not furnish much of an explanation for why one practice or technology was more successful than another.

Of course models of technological evolution are no worse off than models of biological evolution in this respect. If we learn that the two-horned ıinos displaced the one-horned rhinos in some population because they vere fitter, we again learn almost nothing about the causes of change. ı both cases, we want to know what made a successful organism or arfact fitter. Here, while biological explanations might have recourse to natomy, physiology, or ecology, techno-evolutionary explanations will ave recourse to the traditional human sciences of psychology, sociolgy, anthropology, and economics. If we are told that wearing baseball aps backwards is fitter than wearing them the right way around in some ociety, then we still want to ask why this cap orientation is fitter. And ere some quite standard sociological explanation may prevail in terms f status, or sexual attractiveness, or the assertion of group membership. ven classical economic models will have a role; the price of goods will ften be a determinant of their fitness, as will their quality. Once these xplanations are fleshed out they will not look so different from what ve standardly think of as economic, sociological, or psychological explaations for technological change. (See Sober 1992 for a similar line of riticism.)

Older theories of technology change—Marxism, classical economics, reudianism—are thus stronger in some ways than our current best evotionary theories of technology change. Each of these theories is comatible with an evolutionary theory, and can be understood as offering useful set of principles regarding what the most significant selection ressures working on artifacts are likely to be. In terms of what they say bout the determinants of technology change, evolutionary theories do ot add tools to these existing theories—rather, they abstract away from ıem, telling us that technology changes according to some unspecified et of selection pressures.

.7 Looking Ahead

; there nothing of value that an evolutionary approach can bring? In the nd, questions like this are best answered by example. If enough practioners of an evolutionary theory of technology change are able to gain luminating insights that have thus far eluded investigators from more traitional schools, then the approach will be vindicated. However, I would ke to gesture at a few areas where an evolutionary approach may give s new understanding.

First, reflection on the nature of evolutionary explanations can hel
to dissolve conflicts between different schools within the humanitie
(economics, psychology, sociology, and others) over which level is appro
priate to explain some cultural change. In chapter 6 we saw that in th
organic case we should be happy to ascribe multiple, nonconflicting func
tions to single traits. Perhaps an enzyme makes some metabolic proces
more efficient, which in turn increases running speed, which in turn help
the organism to evade predators, which in turn helps it to leave offspring
All of these can be admitted as functions of the enzyme. If our inter
est is instead to explain the fitness of some artifact, then we can cite
variety of different functions in just the same way. So we might speak o
the physiological effects it has, which in turn might cause psychologica
effects, which in turn might lead to broad socioeconomic effects, which i
turn may result in the artifact type increasing its representation. The evo
lutionary approach provides a way of reconciling apparently conflictin
accounts of the functions of artifacts that come from different sciences.

As well as distinguishing different functions for an artifact in terms o
different downstream effects it has within a single selection process, th
evolutionary account also allows us to distinguish different levels of selec
tion. So we might point to a number of different selection processes tha
an artifact has to go through, and we can ask for different functions i
each. Hence, traits of the artifact that look like “junk” from the perspec
tive of the selective demands of its marketplace may have a clear functio
in helping the artifact through the idiosyncrasies of a corporate budgetin
process.

Although selection explanations are usually thought to explain goo
design, here their primary use may be to explain poor design with respec
to one selective regime, by reference to good design for a different regime
Many R&D processes can be understood as nested selection processe
An early idea for an airfoil is tested in the mind of a designer agains
some imaginary demands. Then a model is made, and the airfoil is teste
in a wind tunnel. Finally, the airfoil is produced in bulk and marketed t
plane manufacturers. Whether or not these selection processes take plac
on models in a wind tunnel, on diagrams drawn on a piece of paper, o
between imagined prototypes in the mind of the inventor, the selectio
process itself only makes likely the emergence of the fittest variant fo

e environment in which it finds itself. So we should expect models well
uited to the demands of wind tunnels (and the people who manage them),
iagrams well suited to the demands of those studying the diagrams, and
epresentations well suited to the imagined criteria of the inventor. The
ems which emerge from these processes will be well suited to real-world
tuations only if the selective environment also succeeds in represent-
g the demands of the external world, and if the imagined or modeled
teractions between the models and the modeled environment also re-
ect real interactions between the external environment and the ultimate
sed artifact. The discovery of selection processes—what Vincenti (1990,
000) calls "vicarious selectors"—within all types of corporation, and
eir alignment with the demands of the marketplace, could constitute
n important set of steps for enterprises keen to match their output to
emand.

Finally, and most interestingly, we need to remember the lessons learned
chapter 2 about the relationship between evolution and development.
Ve saw that selection will not result in good design unless the develop-
ental organization of an organism is of a certain type. What does this
ean in the case of artifacts? The developmental program of a species
onstrains the likely possibilities for morphological change open to it at a
oment in time. Similarly, when we think of the developmental program
f an artifact, this concept needs to reflect the likely possibilities for how
n artifact's form might change at a moment in time. So this program needs
be interpreted very broadly to include parameters for change fixed by
ll of the processes that go into its formation. These will include factors
elating to how the artifact is conceived by its designer, or, in the case of
corporate research effort, by the group of designers that fashion it. The
aim that only those artifacts with the right developmental organization
ill evolve to show complex adaptations is supported by the common
bservations that to solve a problem one must learn how to think about
in the right way, how to organize parts so that functional subsystems do
ot interact with each other in detrimental ways, how to represent design
arameters in preliminary drawings, how to measure performance and
forth.

The evolutionary perspective thereby gives some kind of hint at what
ight separate creative geniuses from the rest of us, and gives some hope

that future work in evolutionary developmental biology might help us
flesh out these hints. Those who have creative success bring a set of sele
tion pressures to bear on prototype artifacts that mimic selection pressur
in broader marketplaces—that is why the items they produce are subs
quently so successful. Yet they also approach the problem in the rig
way—their modes of splitting problems into subproblems, and introdu
ing mutation into the selected artifacts are apt for cumulative evolutio
Design success is the result of the right developmental organization f
artifacts as much as a result of the right selection pressures.

7.8 Intention and Artifact Evolution

I have accepted a number of disanalogies, or at least differences in degre
between features of artifact evolution and organic evolution. I have n
yet addressed what may seem to be a central point of disanalogy—name
the fact that technological evolution is an intentional, goal-directed pr
cess driven by desires for new types of artifact, beliefs about how artifac
work, and so forth. I argued in section 7.3 that the intelligence of huma
designers is responsible for the large number of potential replicators
technological evolution, and that this does make the application of sta
dard evolutionary models to technology change difficult. Yet this does n
address the thought that the fact that human designers aim toward som
goal is what makes the important difference. Does it?

One might think that intelligence obviates the need for a selection h
pothesis to explain good design. Selection theories explain design whe
there are no plans and no intentions. When we have real designers, the o
jection goes, with real intentions and real plans, no such theory is neede
This would be a poor objection to applying selection theories in the tec
nological domain because it is not clear how even real intentions and re
plans explain the emergence of good design. It is certainly not enoug
to explain how an excellent watch comes into existence to say that th
designer intended to make an artifact that would tell the time. Poor d
signers and good designers alike have intentions and make plans, yet on
some of these result in good design.

Is selection needed to explain the good design of artifacts? Might som
other explanation such as "creative genius" or "insight" offer a bett

explanation for the successes of designers, architects, or composers? By way of response, we should say two things. First, to offer "genius" as an explanation for creative success is really to offer no explanation at all. The goal of understanding creativity is to explain how it is that some of us who want to produce wonderfully engineered artifacts or perfectly crafted music are unable to carry out these desires, while a few people are. To label these few with the power of "creative genius" is simply to rename the problem. Second, to the extent that we think of this process as explicable at all, evolutionary developmental processes look to be quite promising explanations of success—at least for some artificers at some times. There is a good deal of anecdotal evidence to suggest that the process of invention follows an algorithm where a set of variants is created and some are selected for further modification. In an investigation of Edison's invention of the telephone by Carlson and others (Carlson et al. 2000) the authors show how Edison's sketches seem to be traceable into fairly coherent lineages, and how they involve modifications of various sorts which can be thought of as competing with each other for representation in future generations of research and redesign.

This said, we should resist the thought that all explanations for design success need be selection explanations. We have already seen two alternative explanations. Some good design, as in the case of combinatorial chemistry, may be the result of a simple sorting process. The further modification of successful variants may not always apply; in some cases good design may be achieved simply by running through a large number of alternatives and stopping when a good solution is found. What is more, in some cases good design is brought about through knowledge and deduction, and in the absence of variation, parts with well-understood functions are brought together in a way that the artificer knows will bring about the intended effect. In such cases, there is no selection or sorting process to explain design at a proximal level. Perhaps at a more distal level selection processes underlie these cases of innovation also. The complex screening technologies of the pharmaceutical industry do not fall from the sky, and a creature that can gather items of knowledge and put them together in novel ways is a great evolutionary achievement. Yet if what we are interested in is proximal explanations for design success we need not always invoke Darwinian mechanisms.

7.9 Intelligent Design Creationism

In many cases, artifacts owe their good design to the action of a selection process. That shows that unless intelligent design is further disambiguated, it cannot stand as a genuine alternative to a selection theory in the explanation of good design. And that, in turn, puts certain limitations on the pretensions of intelligent design creationists, who need to contrast explanations for the design of organisms that look to selection with explanations that look to intelligence.

A particularly forceful way of presenting this problem arises out of the work of William Dembski. In his book *The Design Inference,* Dembski (1998) asks us to consider three types of explanation for the appearance of patterned states in nature. They are regularity, chance, and design. If we are presented with some pattern of markings on a cave wall, say, we might think of these as produced by regularity (the mineral that forms the wall always crystallizes in that pattern), or perhaps the pattern owes itself to chance (water erosion just happens to have worn that pattern), or perhaps we should instead attribute the pattern to design (the pattern was carved by early man). Fitelson et al. (2001) have already done an excellent job of undermining Dembski's version of the design inference, but we can now add to their critique by pointing out that in many cases the way in which agents produce good designs is through a selection process.

This leaves Dembski with a dilemma. His "explanatory filter" tells us that if we can reject regularity and chance as explanations for some pattern, then we should infer that design is responsible. But if "design" is defined so as to include selection processes, then his design inference cannot show that intelligence, rather than a selection process alone, is responsible for patterns we conclude to be "designed." In fact, Dembski himself does not want to define his concept of "design" in terms of intelligence. Rather, his definition is negative: "Defining design as the negation of regularity and chance avoids prejudicing the causal stories we associate with design inferences" (1998, p. 36). This move founders on the other horn of the dilemma, for it is hard to see what kind of a process design could be if it cannot involve regularity or chance. After all, selection—even selection that goes on in the mind of an agent—involves both regularity and chance. So Dembski's definition of design now seems to exclude the kinds of selection processes that have produced many of the artifacts we use.

The proponent of any form of intelligent design creationism needs to tell
s why intelligence can explain something that a nonintelligent selection
rocess cannot. More particularly, intelligent design needs to be charac-
rized in such a way as not to include selection processes as the means by
hich such design is achieved. If the hypothesized designer achieves an
xcellent design for the universe, or some part of it, using an internalized
lection process, we can surely ask why we may not posit as a better
xplanation for the design in question that same selection process acting
lone, without the gratuitous addition of an intelligent agent. Suppose, for
xample, we agree that the adaptedness of Planck's constant to the exis-
nce of life is something that stands in need of explanation. This first step
the argument is already contentious. But how, exactly, is an intelligent
esigner supposed to have discovered this suitable value? If the answer is
In roughly the same way as human designers—through a kind of selec-
on process," then on the face of it we could simply posit some kind of
lind selective regime that achieves the same end, without the superfluous
ddition of agency (that is roughly what Smolin 1997 proposes). If the
nswer is "It's a mystery how the designer worked it out," then it is better
remain content, as Hume did, with the mystery of adaptation, rather
an introducing an intelligent designer who designs through mysterious
eans. The proponent of intelligent design creationism faces the task not
erely of telling us why certain facts truly require explanations, but we
lso need to know, in some detail, how the designer goes about discover-
g the principles of good design that the facts are supposed to embody.
Without this second part of the story we have no reason to think that
telligent agency, rather than a mere selection process, is truly required
explain the perplexing facts in question.

Dembski himself (2001) tries to take on these arguments. The property
f "good design" for Dembski is understood to be that of instantiating
complex specified information" (CSI). Consider the text of Vikram Seth's
he Golden Gate. The claim that it shows good design can be cashed out
y saying, first, that it is a highly unlikely arrangement of letters to have
een produced by a random text generator. Yet we would not say that
n equally long string of senseless text was well designed, in spite of the
act that it is equally unlikely to have been produced by a random text
enerator, and in spite of the consequent fact that, in Dembski's sense, it
ontains complex information. That is why good design needs to meet the

further condition of *specification*. What Dembski says about specificatic is difficult, but an example will give us a flavor of the idea. If an arrow fired at a wall and hits the target, this gives us evidence for the skill of th archer. If the arrow hits a wall and a target is drawn around the arro this gives us no such evidence. The first outcome is *specified*, the se ond *unspecified*. The specified patterns are the kinds of patterns who actualization Dembski thinks provides evidence of design.

In fact, there is nothing inherent to the patterns themselves that mak us infer agency in these cases; it is our background knowledge about th phenomena in question. It is because we know that meaningful symbo (and not long strings of nonsense) are the kinds of things that autho strive to produce that meaningfulness in a symbol string is evidence i favor of design; it is because we know that archers strive to hit predraw targets (and not just any part of a surface) that hitting such a target evidence of skill in attaining such a goal on the part of the archer.

That way of characterizing how we infer design looks likely to cu Dembski's own design inference off short. It invites us to settle the que tion of whether eyes are produced by natural selection or intelligence b looking at whether an explanation of their form that looks to intelligen is better, in the light of our background information, than an explanatic that looks to natural selection. Since we know that eyes greatly promo an organism's fitness, and since we have evidence that selection is able t fashion traits that promote an organism's fitness, it seems that selection a good explanation for the emergence of an eye. Maybe the creationist ca show us why this explanation is lacking, but it's hard to see why definin *specification* needs to feature in the exchange.

We can nevertheless hazard a generic characterization of specificatio that makes both eyes and books specified, and that does not lead to trivial resolution in favor either of the creationist camp or of the evc lutionist camp. So let us say that a pattern is specified if (but not onl if) it is suitable to a set of outcomes beneficial either to agents or to o ganisms. It is because *The Golden Gate* is useful for someone wantin to read an excellent book that this unlikely pattern suggests design t us, while unlikely strings of nonsense do not. This definition captur what creationists and evolutionists agree on about eyes—that they ar good for organisms. Since creationists and evolutionists also agree th

ıe structure of the vertebrate eye is unlikely to arise through a single mu-
ıtion, both camps will agree that eyes contain a good deal of information.
yes, like books, carry complex specified information on this definition.
ut Dembski gives us no argument at all for why "pure chance, entirely
nsupplemented and left to its own devices, is incapable of generating
:SI" (2001, p. 570). "Pure chance" here is supposed to mean any con-
ngent process that does not involve intelligent design. Dembski owes us
n account of what it wrong with the evolutionist's explanation for how
election—a "pure chance" explanation in this sense—can generate CSI.
ınd since the way in which intelligence itself brings CSI into existence is
ften through a selection process, it must be the case that pure chance
sometimes able to generate CSI.

Now Dembski will doubtless reply that I have grossly oversimplified
ıis account of specification—indeed, I have. Yet it is very hard to see
ıow any augmented definition of specification will do what the intelligent
esign creationist would need. It would need to have the consequence that
yes, universal constants, or whatever were specified, while also showing
hat their form could not be explained by a selection process. Consider
)embski's (2001, p. 562) generalization of the case where the arrow hits
prepainted target. He claims that if a pattern "is given prior to the
ossibility being actualized . . . then the pattern is automatically specified."
urely organic environments give patterns for organic forms prior to those
orms being actualized, in roughly the same way that a target painted
rior to an arrow's flight gives a pattern for that arrow's position prior
o that position being actualized. If adaptations are indeed instances of
:SI on this definition, then we have seen no reason to doubt that natural
election can explain CSI. If adaptations are not instances of CSI on this
lefinition, then no one can use them to argue for intelligent design.

'.10 Conclusions

t is time to return to the basic phenomena of artifact talk in biology that
ıave been the central topic of this book. If both organisms and artifacts
ıre often subject to selection processes, and if artifacts themselves can
herefore be thought of as having (technological) evolutionary functions,
:hen does that show that these evolutionary functions are more basic than

intended functions? That would be an idle claim. Intended functions and evolutionary functions are different. To review some of the reasons we have seen for this: explanations in terms of intention often explain not the makeup of an artifact but the actions of the artificer; intentional actions can affect individual objects, while evolutionary functions refer to population-level phenomena; many artifacts have evolutionary functions that are not intended functions, for these artifacts have effects that augment their fitness even though these effects are unintended; many artifacts have intended functions that are not evolutionary functions, for the intentions of the artificer mistake the selection pressures at work on the items, or because the intention of the artificer in some particular token case does not reflect the general selection pressures at work on artifacts of the type. What is more, we have seen that in some cases artifacts might be best viewed not from the perspective of a selection process, but from the perspective of a simpler sorting process.

Still, does the fact that artifact functions can at least sometimes be thought of as evolutionary functions establish finally that the kinds of functions organisms have must themselves be thought of as full-blooded "genuine" functions, and not mere "as-if" functions? Again, that is an idle claim. What goes for organisms goes for artifacts. We now understand the difference between an intended artifact function and an evolutionary artifact function. And evolutionary functions do not differ in their claims to "real" teleology from sorted functions, the functions of stones on the beach. Now that we understand how these concepts work, what their roles are, and how they relate to each other, there are no resources left to enable us to label one "genuine" and another "fake." Nor is there any interest in doing so. Our work, for the moment, is done.

References

Allen, C., and Bekoff, M. (1995). "Biological Function, Adaptation, and Natural Design." *Philosophy of Science* 62: 609–622.

Allen, C., Bekoff, M., and Lauder, G. V., eds. (1998). *Nature's Purposes*. Cambridge, Mass.: MIT Press.

Amundson, R. (1989). "The Trials and Tribulations of Selectionist Explanations," in K. Hahlweg and C. Hooker, eds., *Issues in Evolutionary Epistemology*. New York: SUNY Press, pp. 413–432.

———. (2001). "Adaptation and Development: On the Lack of Common Ground," in S. Orzack and E. Sober, eds., *Optimality and Adaptationism*. Cambridge: Cambridge University Press.

Amundson, R., and Lauder, G. V. (1994). "Function without Purpose: The Uses of Causal Role Function in Evolutionary Biology." *Biology and Philosophy* 9: 443–469.

Ariew, A., Cummins, R., and Perlman, M., eds. (2002). *Functions: New Essays in the Philosophy of Psychology and Biology*. Oxford: Oxford University Press.

Aunger, B., ed. (2000). *Darwinizing Culture: The Status of Memetics as a Science*. Oxford: Oxford University Press.

———. (2002). *The Electric Meme: A New Theory of How We Think*. New York: The Free Press.

Ayala, F. (1970). "Teleological Explanations in Evolutionary Biology." *Philosophy of Science* 37: 1–15.

Barkow, J., Cosmides, L., and Tooby, J., eds. (1992). *The Adapted Mind: Evolutionary Psychology and the Generation of Culture*. Oxford: Oxford University Press.

Basalla, G. (1988). *The Evolution of Technology*. Cambridge: Cambridge University Press.

Baxandall, M. (1985). *Patterns of Intention*. London: Yale University Press.

Beatty, J. (1980). "Optimal Design Models and the Strategy of Model Building in Evolutionary Biology." *Philosophy of Science* 47: 532–561.

Bedau, M. (1991). "Can Biological Teleology be Naturalised?" *Journal of Philosophy* 88: 647–657.

Bell, C. (1837). *The Hand: Its Mechanism and Vital Endowments as Evincing Design*. London: William Pickering.

Bigelow, J., and Pargetter, R. (1987). "Functions." *Journal of Philosophy* 84: 181–196.

Blackmore, S. (1999). *The Meme Machine*. Oxford: Oxford University Press.

Boorse, C. (1976). "Wright on Functions." *Philosophical Review* 85: 70–86.

———. (2002). "A Rebuttal on Functions," in Ariew, A., Cummins, R., and Perlman, M. eds., *Functions: New Essays in the Philosophy of Psychology and Biology*. Oxford: Oxford University Press.

Brakefield, P. (1987). "Industrial Melanism: Do We Have the Answers?" *Trends in Ecology and Evolution* 2: 117–122.

Brandon, R. (1990). *Adaptation and Environment*. Princeton: Princeton University Press.

Buller, D. (1998). "Etiological Theories of Function: A Geographical Survey." *Biology and Philosophy* 13: 505–527.

———, ed. (1999). *Function, Selection, and Design*. New York: SUNY Press.

Campbell, D. C. (1979). "Comments on the Sociobiology of Ethics and Moralising." *Behavioral Science* 24: 37–45.

Cairns-Smith, A. G. (1982). *Genetic Takeover*. Cambridge: Cambridge University Press.

Carlson, W. B. (2000). "Invention and Evolution: The Case of Edison's Sketches of the Telephone," in J. Ziman, ed., *Technological Innovation as an Evolutionary Process*. Cambridge: Cambridge University Press.

Charlesworth, B., and J. T. Giesel (1972). "Selection in Populations with Overlapping Generations, II. Relations Between Gene Frequency Change and Demographic Variables." *American Naturalist* 106: 388–401.

Connor, S. (2001). "Why the Currents in Muesli Put Nuts at the Top." *Independent*, November 15, 2001.

Cosmides, L. (1989). "The Logic of Social Exchange: Has Natural Selection Shaped how Humans Reason? Studies with the Wason Selection Task." *Cognition* 31, 187–276.

Cosmides, L., and Tooby, J. (1992). "The Psychological Foundations of Culture," in J. Barkow, L. Cosmides, and J. Tooby, eds., *The Adapted Mind: Evolutionary Psychology and the Generation of Culture*. Oxford: Oxford University Press.

Cosmides, L., and Tooby, J. (1997). Letter to the Editor of *The New York Review of Books* on Stephen Jay Gould's "Darwinian Fundamentalism" (June 12, 1997) and "Evolution: The Pleasures of Pluralism" (June 26, 1997). Available at http://cogweb.ucla.edu/Debate/CEP_Gould.html. Accessed July 7, 2003.

Creed, E., Lees, D., and Bulmer, M. (1980). "Pre-Adult Viability Differences of Melanic *Biston Betularia.*" *Biological Journal of the Linnean Society* 13: 251–262.

Crow, J. F. (1979). "Genes that Violate Mendel's Rules." *Scientific American* 240(2): 134–146.

Cummins, R. (1975). "Functional Analysis." *Journal of Philosophy* 72: 741–764.

Cziko, G. (1995). *Without Miracles: Universal Selection Theory and the Second Darwinian Revolution.* Cambridge, Mass.: MIT Press.

Darden, L., and Cain, J. (1989). "Selection Type Theories." *Philosophy of Science* 56: 106–129.

Darwin, C. (1996). *The Origin of Species.* Oxford: Oxford University Press. (First Published 1858.)

Davidson, D. (1978). "What Metaphors Mean." *Critical Inquiry* 5: 31–47.

Davies, P. S. (2000a). "Malfunction." *Biology and Philosophy* 15: 19–38.

———. (2000b). "The Nature of Natural Norms: Why Selected Functions are Systemic Capacity Functions." *Noûs* 34: 85–108.

———. (2001). *Norms of Nature: Naturalism and the Nature of Functions.* Cambridge, Mass.: MIT Press.

Dawkins, R. (1976). *The Selfish Gene.* Oxford: Oxford University Press.

———. (1982). *The Extended Phenotype.* Oxford: Oxford University Press.

———. (1986). *The Blind Watchmaker.* New York: Norton.

Dembski, W. (1998). *The Design Inference: Eliminating Chance Through Small Probabilities.* Cambridge: Cambridge University Press.

———. (2001). "Intelligent Design as a Theory of Information," in Pennock, ed., *Intelligent Design Creationism and Its Critics.* Cambridge, Mass.: MIT Press.

Dennett, D. C. (1983). "Intentional Systems in Cognitive Ethology: The 'Panglossian Paradigm' Defended." *Behavioral and Brain Sciences* 6: 343–390.

———. (1988). "Précis of The Intentional Stance." *Behavioral and Brain Sciences* 11: 495–546.

———. (1990). "The Interpretation of Texts, People, and Other Artifacts." *Philosophy and Phenomenological Research* 50: 177–194.

———. (1995). *Darwin's Dangerous Idea.* New York: Norton.

Depew, D. J., and Weber, B. H. (1995). *Darwinism Evolving: Systems Dynamics and the Genealogy of Natural Selection.* Cambridge, Mass.: MIT Press.

Derham, W. (1732). *Physico-Theology: Or a Demonstration of the Being and Attributes of God from his Works of Creation.* London: Printed for W. Innys and R. Manby.

Doolittle, W., and Sapienza, C. (1980). "Selfish Genes, the Phenotype Paradigm, and Genome Evolution." *Nature* 284: 601–603.

Drickamer, L., and Vessey, S. H. (1992). *Animal Behavior: Mechanisms, Ecology and Evolution.* Dubuque, Iowa: Wm. C. Brown.

Fleck, J. (2000). "Artefact↔Activity: The Coevolution of Artefacts, Knowledge and Organization in Technological Evolution," in J. Ziman, ed., *Technological Innovation as an Evolutionary Process.* Cambridge: Cambridge University Press.

Fisher, R. A. (1958). *The Genetical Theory of Natural Selection.* Second Edition. New York: Dover. (First published 1930.)

Fitelson, B., Stephens, C., and Sober, E. (2001). "How Not to Detect Design," in Pennock, ed., *Intelligent Design Creationism and Its Critics.* Cambridge, Mass.: MIT Press.

Ghiselin, M. (1983). "Lloyd Morgan's Canon in Evolutionary Context." *Behavioral and Brain Sciences* 6: 362–363.

Godfrey-Smith, P. (1993). "Functions: Consensus Without Unity." *Pacific Philosophical Quarterly* 74: 196–208.

———. (1994). "A Modern History Theory of Functions." *Nous* 28: 344–362.

———. (1996). *Complexity and the Function of Mind in Nature.* Cambridge: Cambridge University Press.

———. (2001). "Three Kinds of Adaptationism," in S. Orzack and E. Sober, eds., *Optimality and Adaptationism.* Cambridge: Cambridge University Press.

Goode, R., and Griffiths, P. (1995). "The Misuse of Sober's Selection for/Selection of Distinction." *Biology and Philosophy* 10: 99–108.

Goodwin, B. (1994). *How the Leopard Changed Its Spots: The Evolution of Complexity.* New York: Charles Scribner and Sons.

Gordon, D. M., Paul, R. E., and Thorpe, K. (1993). "What Is the Function of Encounter Patterns in Ant Colonies?" *Animal Behaviour* 45: 1083–1100.

Gould, S. J. (1980). *The Panda's Thumb.* New York: Norton.

Gould, S. J., and Lewontin, R. (1979). "The Spandrels of San Marco and the Panglossian Paradigm: A Critique of the Adaptationist Programme." *Proceedings of the Royal Society,* vol. B205: 581–598.

Gould, S. J., and Vrba, E. (1982). "Exaptation—A Missing Term in the Science of Form." *Paleobiology* 8: 4–15.

Gray, R. (1992). "Death of the Gene: Developmental Systems Strike Back," in P. E. Griffiths, ed., *Trees of Life: Essays in the Philosophy of Biology.* Boston: Kluwer: 165–209.

Griffiths, P. (1993). "Functional Analysis and Proper Function." *British Journal for the Philosophy of Science* 44: 409–422.

———. (1996). "The Historical Turn in the Study of Adaptation." *British Journal for the Philosophy of Science* 47: 511–532.

Griffiths, P., and Gray, R. (1994). "Developmental Systems and Evolutionary Explanation." *Journal of Philosophy* 91: 277–304.

Ho, M.-W., and Saunders, P. T. (1984). "Pluralism and Convergence in Evolutionary Theory," in M.-W. Ho and P. T. Saunders, eds., *Beyond Neo-Darwinism: An Introduction to the New Evolutionary Paradigm.* London: Academic.

Howlett and Majerus, M. E. N. (1987). "The Understanding of Industrial Melanism in the Peppered Moth." *Biological Journal of the Linnean Society* 30: 31–44.

Hull, D. (1988). *Science as a Process.* Chicago: University of Chicago Press.

———. (1998). "Introduction to Part IV," in D. Hull and M. Ruse, eds., *The Philosophy of Biology.* Oxford: Oxford University Press.

Jablonka, E. (2000). "Lamarckian Inheritance Systems in Biology: A Source of Metaphors and Models in Technological Evolution," in J. Ziman, ed., *Technological Innovation as an Evolutionary Process.* Cambridge: Cambridge University Press.

Jablonka, E., and Lamb, M. (1995). *Epigenetic Inheritance and Evolution: The Lamarckian Dimension.* Oxford: Oxford University Press.

Jablonka, E., and Szathmary, E. (1995). "The Evolution of Information Storage and Heredity." *Trends in Ecology and Evolution* 10(5): 206–211.

Jardine, N. (1967). "The Concept of Homology in Biology." *British Journal for the Philosophy of Science* 18: 125–139.

Jardine, N., and Sibson, R. (1971). *Mathematical Taxonomy.* Chichester: Wiley.

Kant, I. (1952). *Critique of Judgement* (translated by James Creed Meredith). Oxford: Oxford University Press.

Kauffman, S. A. (1993). *The Origins of Order: Self-Organization and Selection in Evolution.* New York: Oxford University Press.

———. (1995). *At Home in the Universe.* New York: Oxford University Press.

Kettlewell, H. B. R. (1973). *The Evolution of Melanism.* Oxford: Oxford University Press.

Kingsolver, J., and Koehl, M. (1985). "Aerodynamics, Thermoregulation, and the Evolution of Insect Wings: Differential Scaling and Evolutionary Change." *Evolution* 39: 488–504.

Kimura, M. (1991). "Some Recent Data Supporting the Neutral Theory," in M. Kimura and N. Takahata, eds., *New Aspects of the Genetics of Molecular Evolution.* Japan Scientific Societies Press/Springer Verlag: Berlin.

Kitcher, P. (1993). "Function and Design." *Midwest Studies in Philosophy* 18: 379–397.

Krebs, J., and Davies, N. (1997). "Introduction to Part Two," in J. Krebs and N. Davies, eds., *Behavioural Ecology: An Evolutionary Approach.* Fourth Edition. Oxford: Blackwell Science.

Lack, D. (1968). *Ecological Adaptations for Breeding in Birds.* London: Methuen.

Lauder, G. V. (1994). "Homology, Form, and Function," in B. K. Hall, ed., *Homology: The Hierarchical Basis of Comparative Biology.* San Diego: Academic Press.

———. (1996). "The Argument from Design," in M. Rose and G. Lauder, eds., *Adaptation.* Academic Press: San Diego.

Lewens, T. M. (forthcoming-a). "Prospects for Evolutionary Policy." *Philosophy.*

———. (forthcoming-b). "Seven Types of Adaptationism," in D. M. Walsh, ed., *Twenty-Five Years of Spandrels.*

Lewontin, R. (1970). "The Units of Selection." *Annual Review of Ecology and Systematics* 1: 1–18.

———. (1978). "Adaptation." *Scientific American* 239(3): 212–230.

———. (1984). "Adaptation," in Sober, E., ed., *Conceptual Issues in Evolutionary Biology* (first edition). Cambridge, Mass.: MIT Press.

———. (1985). "The Organism as Subject and Object of Evolution," in R. Levins and R. Lewontin *The Dialectical Biologist.* Cambridge, Mass.: Harvard University Press.

Lipton, P. (1991). *Inference to the Best Explanation.* London: Routledge.

Lloyd, L. (1995). "Unit of Selection," in E. F. Keller and L. Lloyd, eds., *Keywords in Evolutionary Biology.* Cambridge, Mass.: Harvard University Press.

Love, M. (1980). "The Alien Strategy." *Natural History* 89(5): 30–32.

Majerus, M. E. N. (1998). *Melanism: Evolution in Action.* Oxford: Oxford University Press.

Manning, R. N. (1997). "Biological Function, Selection, and Reduction." *British Journal for the Philosophy of Science* 48: 69–82.

Matthen, M. (1997). "Teleology and the Product Analogy." *Australasian Journal of Philosophy* 75: 21–37.

Matthen, M., and Ariew, A. (2002). "Two Ways of Thinking About Selection." *Journal of Philosophy* 49: 53–83.

Maynard Smith, J. (1969). "The Status of Neo-Darwinism," in Waddington, ed., *Towards a Theoretical Biology.* Edinburgh: University Press.

———. (1978). "Optimization Theory in Evolution." *Annual Review of Ecology and Systematics* 9: 31–56.

Maynard Smith, J., and Savage, R. (1956). "Some Locomotory Adaptations in Mammals." *Zoological Journal of the Linnean Society* 42: 603–622.

Maynard Smith, J., and Szathmary, E. (1995). *The Major Transitions in Evolution* Oxford: W. H. Freeman/Spektrum.

Mayr, E. (1961). "Cause and Effect in Biology." *Science* 134: 1501–1506.

———. (1963). *Animal Species and Evolution.* Cambridge, Mass.: Harvard University Press.

———. (1974). "Teleological and Teleonomic, A New Analysis," in R. S. Cohen and M. W. Wartofsky, eds., *Methodological and Historical Essays in the Natural and Social Sciences,* vol. 14 of *Boston Studies in the Philosophy of Science.* Dordrecht: D. Reidel.

———. (1982). *The Growth of Biological Thought.* Cambridge, Mass.: Harvard University Press.

———. (1983). "How to Carry Out the Adaptationist Program." *American Naturalist* 121: 324–334.

———. (1993). "Proximate and Ultimate Causation." *Biology and Philosophy* 8: 93–94.

McKitrick, M. (1993). "Phylogenetic Constraint in Evolution: Has it any Explanatory Power?" *Annual Review of Ecology and Systematics* 24: 307–330.

McLaughlin, P. (2001). *What Functions Explain: Functional Explanation and Self-Reproducing Systems.* Cambridge: Cambridge University Press.

Millikan, R. G. (1984). *Language, Truth, and Other Biological Categories.* Cambridge, Mass.: MIT Press.

———. (1989a). "An Ambiguity in the Notion of Function." *Biology and Philosophy* 4: 172–176.

———. (1989b). "Biosemantics." *Journal of Philosophy* 86: 281–297.

———. (1989c). "In Defense of Proper Functions." *Philosophy of Science* 56: 288–302.

———. (1993). *White Queen Psychology and Other Essays for Alice.* Cambridge, Mass.: MIT Press.

Mills, S. K., and Beatty, J. H. (1979). "The Propensity Interpretation of Fitness." *Philosophy of Science* 46: 263–286.

Mokyr, J. (1990). *The Lever of Riches: Technological Creativity and Economic Progress.* Oxford: Oxford University Press.

Moore, J. (1996). *The Death of Competition: Leadership and Strategy in the Age of Business Ecosystems.* Chichester: Wiley.

Nanney, D. L. (1968). "Cortical Patterns in Cellular Morphogenesis." *Science* 160: 496–502.

Neander, K. (1991a). "Functions as Selected Effects: The Conceptual Analyst's Defense." *Philosophy of Science* 58: 168–184.

———. (1991b). "The Teleological Notion of 'Function'." *Australasian Journal of Philosophy* 69: 454–468.

———. (1995a). "Pruning the Tree of Life." *British Journal for the Philosophy of Science* 46: 59–80.

———. (1995b). "Explaining Complex Adaptations: A Reply to Sober's 'Reply to Neander'." *British Journal for the Philosophy of Science* 46: 585–587.

———. (1995c). "Misrepresenting and Malfunctioning." *Philosophical Studies* 79: 109–141.

Nissen, L. (1997). *Teleological Language in the Life Sciences.* Lanham, Maryland: Rowman and Littlefield.

Ollason, J. G. (1987). "Artificial Design in Natural History: Why It's So Easy to Understand Animal Behavior," in P. Bateson and P. Klopfer, eds., *Alternative Perspectives in Ethology*, vol. 7, 233–257. New York: Plenum.

Orgel, L. E., and Crick, F. (1980). "Selfish DNA: The Ultimate Parasite." *Natu* 284.

Orzack, S., and Sober, E. (1994). "Optimality Models and the Test of Adaptationism." *American Naturalist* 143: 361–380.

Oster, G., and Wilson, E. O. (1984). "A Critique of Optimization Theory in Evolutionary Biology," in E. Sober, ed., *Conceptual Issues in Evolutionary Biolog* (First Edition). Cambridge, Mass.: MIT Press.

Over, D. (forthcoming). "The Rationality of Evolutionary Psychology," i J. Bermudez and A. Millar, eds., *Proceeding of the 1998 Stirling Conference o Rationality and Naturalism.*

Oyama, S. (2000a). *The Ontogeny of Information*. Durham, N.C.: Duke University Press.

———. (2000b). *Evolution's Eye*. Durham, N.C.: Duke University Press.

Paley, W. (1802). *Natural Theology or Evidences for the Existence and Attribute of the Deity Collected from the Appearances of Nature*. London: Faulkner.

Papineau, D. (1987). *Reality and Representation*. Oxford: Basil Blackwell.

———. (1993). *Philosophical Naturalism*. Oxford: Basil Blackwell.

Parker, G. A., and Maynard Smith, J. (1990). "Optimality Theory in Evolutionar Biology." *Nature* 348: 27–33.

Pinker, S. (1997). *How the Mind Works*. New York: Norton.

Pinker, S., and Bloom, P. (1990). "Natural Language and Natural Selection." *Behavioral and Brain Sciences* 13: 707–784.

Preston, B. (1998). "Why Is a Wing Like a Spoon? A Pluralist Theory of Functions." *Journal of Philosophy* 95: 215–254.

Quiring, R., Walldorf, U., Kloter, U., and Gehring, W. J. (1994). "Homology of the Eyeless Gene of Drosophila to the Small Eye Gene in Mice and Aniridia in Humans." *Science* 265: 785–789.

Raff, R. (1996). *The Shape of Life: Genes, Development, and the Evolution of Animal Form*. Chicago: University of Chicago Press.

Reeve, H., and P. Sherman (1993). "Adaptation and the Goals of Evolutionary Research." *Quarterly Review of Biology* 68: 1–32.

Reeve, H., and P. Sherman (2001). "Optimality and Phylogeny: A Critique of Current Thought," in S. Orzack and E. Sober, eds., *Optimality and Adaptationism*. Cambridge: Cambridge University Press.

Resnik, D. (1995). "Function Language and Biological Discovery." *Journal for the General Philosophy of Science* 26: 119–134.

———. (1997). "Adaptationism: Hypothesis or Heuristic?" *Biology and Philosophy* 12: 39–50.

Roemer et al. (1997). "Epigenetic Inheritance in the Mouse." *Current Biology* 7: 277–280.

Ruse, M. (1982). "Teleology Redux," in J. Agassi and R. Cohen, eds., *Scientific Philosophy Today: Essays in Honour of Mario Bunge*, vol. 67 of *Boston Studies in the Philosophy of Science*: 299–309. Dordrecht: D. Reidel.

———. (1996). "Booknotes." *Biology and Philosophy* 11: 283–288.

Simon, H. (1996). *The Sciences of the Artificial*. Third Edition. Cambridge, Mass.: MIT Press.

Smith, K. (1992). "Neo-Rationalism Versus Neo-Darwinism: Integrating Development and Evolution." *Biology and Philosophy* 7: 431–451.

Smolin, L. (1997). *The Life of the Cosmos*. Oxford: Oxford University Press.

Sober, E. (1984a). *The Nature of Selection*. Cambridge, Mass.: MIT Press.

———. (1984b). *Conceptual Issues in Evolutionary Biology*. Cambridge, Mass.: MIT Press.

———. (1992). "Models of Cultural Evolution," in P. E. Griffiths, ed., *Trees of Life: Essays in the Philosophy of Biology*. Boston: Kluwer.

———. (1993). *Philosophy of Biology*. Boulder, Colo.: Westview.

———. (1998). "Six Sayings About Adaptationism," in D. Hull and M. Ruse, eds., *The Philosophy of Biology*: 72–86. Oxford: Oxford University Press.

Sperber, D. (2000). "An Objection to the Memetic Approach to Culture," in Aunger, ed., *Darwinizing Culture: The Status of Memetics as a Science*. Oxford: Oxford University Press.

Stephens, D. W., and Krebs, J. (1986). *Foraging Theory*. Princeton: Princeton University Press.

Sterelny, K., and Griffiths, P. (1999). *Sex and Death: An Introduction to the Philosophy of Biology*. Chicago: University of Chicago Press.

Sterelny, K., Smith, K., and Dickison, M. (1996). "The Extended Replicator." *Biology and Philosophy* 11: 377–403.

Symons, D. (1992). "On the Use and Misuse of Darwinism in the Study of Human Behaviour," in J. Barkow, L. Cosmides, and J. Tooby, eds., *The Adapted Mind: Evolutionary Psychology and the Generation of Culture*. Oxford: Oxford University Press.

Thompson, D. W. (1961). *On Growth and Form*. Cambridge: Cambridge University Press. (First published 1917.)

Tinbergen, N. (1963). "On the Aims and Methods of Ethology." *Zeitschrift für Tierpsychologie* 20: 410–433.

Vincenti, W. G. (1990). *What Engineers Know and How They Know It*. Baltimore: The Johns Hopkins University Press.

———. (2000). "Real-World Variation-Selection in the Evolution of Technological Form: Historical Examples," in J. Ziman, ed., *Technological Innovation as an Evolutionary Process*. Cambridge: Cambridge University Press.

Walsh, D. (1996). "Fitness and Function." *British Journal for the Philosophy of Science* 47: 250–264.

———. (2000). "Chasing Shadows." *Studies in History and Philosophy of Biological and Biomedical Sciences* 31: 135–154.

Walsh, D., and Ariew, A. (1996). "A Taxonomy of Functions." *Canadian Journal of Philosophy* 26: 493–514.

Walsh, D., Lewens, T., and Ariew, A. (2002). "The Trials of Life: Natural Selection and Random Drift." *Philosophy of Science* 69: 429–446.

Wehner, R. (1997). "Sensory Systems and Behaviour," in J. Krebs and N. Davies, eds., *Behavioural Ecology: An Evolutionary Approach*. Fourth Edition. Oxford: Blackwell Science.

Wehner, R. *et al.* (1996). "Visual Navigation in Insects: Coupling of Egocentric and Geocentric Information." *Journal of Experimental Biology* 199: 129–140.

Wickler, W. (1976). "Evolution-Oriented Ethology, Kin Selection, and Altruistic Parasites." *Zeitschrift für Tierpsychologie* 42: 206–214.

Williams, G. (1966). *Adaptation and Natural Selection*. Princeton: Princeton University Press.

———. (1992). *Natural Selection: Domains, Levels, and Challenges*. New York: Oxford University Press.

Wright, L. (1973). "Functions." *Philosophical Review* 82: 139–168.

———. (1976). *Teleological Explanation*. Berkeley: University of California Press.

Wright, S. (1932). "The Roles of Mutation, Inbreeding, Crossbreeding, and Selection in Evolution." *Proceedings of the VI International Congress of Genetics*: 356–366.

Ziman, J., ed. (2000). *Technological Innovation as an Evolutionary Process*. Cambridge: Cambridge University Press.

Index